Lecture Notes in Mathematics 1081

David J. Benson

Modular Representation Theory

New Trends and Methods

Second printing

Springer

Author

David J. Benson
Department of Mathematical Sciences
University of Aberdeen
Meston Building
King's College
Aberdeen AB24 3UE
Scotland UK

Library of Congress Cataloging in Publication Data. Benson, David, 1955-. Modular representation theory. (Lecture notes in mathematics; 1081) Bibliography: p. Includes index. 1. Modular representations of groups. 2. Rings (Algebra) I. Title. II. Series: Lecture notes in mathematics (Springer-Verlag); 1081.
QA3.L28 no. 1081 [QA171] 510s [512'.2] 84-20207
ISBN 0-387-13389-5 (U.S.)

Mathematics Subject Classification (1980): 20C20

Second printing 2006

ISSN 0075-8434
ISBN-10 3-540-13389-5 Springer-Verlag Berlin Heidelberg New York
ISBN-13 978-3-540-13389-6 Springer-Verlag Berlin Heidelberg New York

Production: LE-TEX Jelonek, Schmidt & Vöckler GbR, Leipzig
Cover design: *design & production* GmbH, Heidelberg

Printed on acid-free paper SPIN: 11749158 41/3100/YL 5 4 3 2 1 0

Introduction

This book grew out of a graduate course which I gave at Yale University in the spring semester of 1983. The aim of this course was to make some recent results in modular representation theory accessible to an audience ranging from second-year graduate students to established mathematicians.

The material covered has remarkably little overlap with the material currently available in textbook form. The reader new to modular representation theory is therefore encouraged also to read, for example, Feit [51], Curtis and Reiner [37,38], Dornhoff [44], Landrock [65], as well as Brauer's collected works [16], for rather different angles on the subject.

The first of the book's two chapters is intended as background material from the theory of rings and modules. The reader is expected already to be familiar with a large proportion of this, and to refer to the rest as he needs it; proofs are included for the sake of completeness.

The second chapter treats three main topics in detail.

(i) Representation rings.

(ii) Almost split sequences and the Auslander-Reiten quiver.

(iii) Complexity and cohomology varieties.

I hope to impress upon the reader that these three topics are closely connected, and to encourage further investigation of their interplay.

The study of modular representation theory was in some sense started by L. E. Dickson [40] in 1902. However, it was not until R. Brauer [16] started investigating the subject that it really got off the ground. In the years between 1935 and his death in 1977, he almost single-handedly constructed the corpus of what is now regarded as the classical modular representation theory. Brauer's main motivation in studying modular representations was to obtain number theoretic restrictions on the possible behaviour of ordinary character tables, and thereby find restrictions upon the structure of finite groups. His work has been a major tool in the classification of the finite simple groups. For a definitive account of modular representation theory from the Brauer viewpoint (as well as some more modern material) see Feit [51].

It was really J. A. Green who first systematically developed the study of modular representation theory from the point of view of examining the set of indecomposable modules, starting with his paper [54]. Green's results were an indispensable tool in the treatment by Thompson, and then more fully by Dade, of blocks with cyclic defect groups. Since then, many other people have become interested in the study of the modules for their own sake.

In the study of representation theory in characteristic zero, it is customary to work in terms of the character table, namely the square table whose rows are indexed by the ordinary irreducible representations, whose columns are indexed by the conjugacy classes of group elements, and where a typical entry gives the trace of the group element on the representation. Why do we use the trace function? This is because the maps $V \mapsto tr(g,V)$ are precisely the algebra homomorphisms from the representation ring to $\mathbb{C}$, and these homomorphisms separate representations. In particular, in this case the representation ring is semisimple. This has the effect that we can compute with representations easily and effectively in terms of their characters; representations are distinguished by their characters, direct sum corresponds to addition and tensor product corresponds to multiplication. The orthogonality relations state that we may determine the dimension of the space of homomorphisms from one representation to another by taking the inner product of their characters.

How much of this carries over to characteristic p, where $p \mid |G|$? The first problem is that Maschke's theorem no longer holds; a representation may be indecomposable without being irreducible. Thus the concepts of representation ring $A(G)$ and Grothendieck ring do not coincide. The latter is a quotient of the former by the "ideal of short exact sequences" $A_0(G,1)$. Brauer discovered the remarkable fact that the Grothendieck ring $A(G)/A_0(G,1)$ is semisimple, and found the set of algebra homomorphisms from this to $\mathbb{C}$, in terms of lifting eigenvalues. Thus he gets a square character table, giving information about composition factors of modules, but saying nothing about how they are glued together.

In an attempt to generalize this, we define a species of the representation ring to be an algebra homomorphism $A(G) \to \mathbb{C}$. Even if we use the set of all species, we cannot distinguish between modules V_1 and V_2 when $V_1 - V_2$ is nilpotent as an element of $A(G)$. For some time, it was conjectured that $A(G)$ has no nilpotent elements in general. However, it is now known that $A(G)$ has no nilpotent elements whenever kG has finite representation type (i.e. the Sylow p-subgroups of G are cyclic, where $p = \mathrm{char}(k)$), as well as a few other cases in characteristic two, whilst in general there are nilpotents (see O'Reilly [98] and Zemanek [95, 96] as well as a forthcoming paper by J. Carlson and the author).

The Brauer species (i.e. the species of $A(G)$ which vanish on $A_0(G,1)$) may be evaluated by first restricting down to a cyclic subgroup of order coprime to p, and then lifting eigenvalues. The corresponding concept for a general species is the <u>origin</u>, namely the minimal subgroup through which the species factors. We show that the origins of a species are very restricted in shape, namely if H is an origin then $H/O_p(H)$ is a cyclic group of order coprime to p, and we show how $O_p(H)$ is related to the vertices of modules on which the species does not vanish.

Many of the properties of representation rings and species are governed by the <u>trivial source subring</u> $A(G,\mathrm{Triv})$, which is a finite dimensional semisimple subring of $A(G)$. Thus we spend several sections investigating trivial source modules, and showing how these modules are connected with block theory. Instead of developing defect groups and Brauer's first main theorem just for group algebras, we develop them for arbitrary permutation modules, and recover the classical case by applying the theory to $G \times G$ acting on the set of elements of G by left and right multiplication. When applied to this case, the orthogonality relations 2.6.4 become the ordinary character orthogonality relations.

In ordinary character theory, one of the ways in which the structure of the group is reflected in the character table is via the so-called power maps, or Adams operations. Namely there are ring homomorphisms ψ^n on the character ring, with the property that the character value of g on $\psi^n(V)$ is the character value of g^n on V. These are usually given in terms of the exterior power operations Λ^n, and these operations make the character ring into a special lambda-ring. It turns out that for modular representations we must first construct the ring homomorphisms $\psi^n\colon a(G) \to a(G)$, and then use them to construct operations λ^n, which do not agree with the exterior power operations unless $n < p$ (although they do at the level of Brauer characters), and the λ^n make $a(G) \otimes_{\mathbb{Z}} \mathbb{Z}[1/p]$ into a special lambda-ring. It then makes sense to use the ψ^n to define the powers of a species. As an application of these power maps, we give Kervaire's proof that the determinant of the Cartan matrix is a power of p, rather than using Brauer's characterization of characters.

The next feature of ordinary character theory which we may wish to mimic is the fact that the orthogonality relations may be interpreted as saying that a module is characterized by its inner products with the indecomposable modules. It turns out that this is still true in arbitrary characteristic although the proof is much harder. There are two sensible bilinear forms to use here, which both agree with the usual inner product in the case of characteristic zero. These are

$$(V, W) = \dim_k \mathrm{Hom}_{kG}(V, W)$$

and $$\langle V, W\rangle = \text{rank of } \sum_{g \in G} g \text{ on } \mathrm{Hom}_k(V, W).$$

There are elements u and v of $A(G)$ with $uv = 1$, $u^* = v$, $(V, W) = \langle v.V, W\rangle = \langle V, u.W\rangle$ and $\langle V, W\rangle = (u.V, W) = (V, v.W)$. It is thus easy to pass back and forth between the two inner products, and the second has the advantage that it is symmetric. It is not until 2.18, after we have introduced the almost split sequences, that we can prove that these inner products are non-singular on $A(G)$. We do this by finding elements $\tau(V) \in A(G)$, one for each indecomposable module V, such that $\langle V, \tau(W)\rangle$ is non-zero if and only if $V \cong W$. These elements $\tau(V)$ are called <u>atoms</u>, and they are the simple modules and the 'irreducible glues', the latter being related to the almost split sequences. Any module may then be regarded as a formal sum of atoms, namely the composition factors and the glues holding it together.

As an application of the non-singularity theorem, in 2.19 we find the radical of the bilinear forms $\dim_k \mathrm{Ext}^n_{kG}(-\,,\,-)$ on $A(G)$.

In section 2.21 we bring together these results on representation rings to provide an extension of Brauer character theory. We project all the information we have onto a finite dimensional direct summand of $A(G)$ satisfying certain natural conditions. We define tables T_{ij} and U_{ij} called the <u>atom table</u> and <u>representation table</u>, which satisfy certain orthogonality relations. The minimal direct summand satisfying our conditions is the summand spanned by the projective modules. In this case T_{ij} is the table of Brauer characters of irreducible modules, and U_{ij} is the table of Brauer characters of projective indecomposable modules. The analogues in the general case for the centralizer orders in the Brauer theory are certain algebraic numbers which need not be positive or rational.

We have now reached a position where we would like to understand better how to compute inside $A(G)$. This means that we need to understand the behaviour of tensor products of modules. One of the most interesting tools we have available for this at the moment is Carlson's idea of associating varieties to modules. To each module we associate a homogeneous subvariety of $\mathrm{Spec}\, H^{ev}(G,k)$, the spectrum of the even cohomology ring of G. These varieties $X_G(V)$ have the properties that $X_G(V \oplus W) = X_G(V) \cup X_G(W)$, $X_G(V \otimes W) = X_G(V) \cap X_G(W)$, and if V is indecomposable then the projective variety $\overline{X}_G(V)$ corresponding to $X_G(V)$ is connected. Thus at the level of representation rings, if X is a homogeneous subset of $\mathrm{Spec}\, H^{ev}(G,k)$, then the linear span $A(G,X)$ of the modules V with $X_G(V) \subseteq X$ is an ideal in $A(G)$.

Generalizing a result of Quillen, Avrunin and Scott [9] have shown how to stratify $X_G(V)$ into strata corresponding to restrictions of V to elementary abelian p-subgroups E of G. Thus many properties of modules are controlled by these restrictions. For example, Chouinard's theorem states that V is projective if and only if $V{\downarrow}_E$ is projective for all such E.

The dimension of $X_G(V)$ is an important invariant of V, called the <u>complexity</u>, $cx_G(V)$. It measures the rate of growth of a minimal projective resolution of V, and the Alperin-Evens theorem states that $cx_G(V)$ is equal to the maximal complexity of $V{\downarrow}_E$ as E ranges over the elementary abelian p-subgroups of G.

All the results in this area seem to depend on two basic results, namely a theorem of Serre (2.23.3) on products of Bocksteins for p-groups, and the Quillen-Venkov lemma (2.23.4). In order to prove and use these results, we need to introduce the Lyndon-Hochschild-Serre spectral sequence, and the Steenrod algebra. The former is introduced in section 2.22 without complete proofs, while the latter is introduced at the beginning of 2.23 with no proofs at all! This is because complete construction of these tools would take us too far away from the purpose of this book. However, we do give a very sketchy outline of the construction of the Steenrod operations in characteristic two in an exercise at the end of 2.23.

We return to the almost split sequences in sections 2.28 to 2.33, and show how these may be fitted together to form a locally finite graph (the <u>Auslander-Reiten quiver</u>). We pull off the projective modules, since these often get in the way, and are then left with the <u>stable quiver</u>. This is an example of an abstract <u>stable representation quiver</u>, and the Riedtmann structure theorem (2.29.6) shows that a connected stable representation quiver is uniquely expressible as a quotient of the universal covering quiver of a tree, by an 'admissible' group of automorphisms. This tree is called the <u>tree class</u> of the connected component of the stable quiver. Using an invariant $\eta(V)$ related to the complexity of V, together with the finite generation of cohomology, we prove Webb's theorem (2.31.2), that for modules for a group algebra, the tree class of a connected component is either a Dynkin diagram (finite or infinite) or a Euclidean diagram. Following Webb, we investigate each of these possibilities in turn, and the results are summarized in 2.32.6.

One surprising corollary of Webb's theorem is given in 2.31.5, which states that if P is a non-simple projective indecomposable kG-module, then Rad(P)/Soc(P) falls into at most four indecomposable direct summands. In practice, this module will usually be indecomposable. For the alternating group A_4 over the field of two elements, there is a projective indecomposable module P such that Rad(P)/Soc(P) has three summands, but I know of no examples with four.

We include exercises at the end of most sections. These vary substantially in difficulty, from routine exercises designed to familiarize the reader with the concepts of the text, to outlines of recent related results for which there was not enough room in the text.

We also include an appendix containing descriptions of the representation theory of some particular groups, in order to illustrate some of the concepts in the text and provide the reader with concrete examples.

Finally, it is my pleasure to thank all those who made this book possible. I would particularly like to thank Richard Parker, Peter Landrock, Jon Carlson, Peter Webb and Walter Feit for sharing their insights with me, Yale University for employing me during the period in which I was writing this book, and Mel DelVecchio for her patience in typing the manuscript.

Table of Contents

Section One: Rings and Modules

Section Two: Modules for Group Algebras

Conventions and Abbreviations

All rings have an identity element, although homomorphisms need not take identity elements to identity elements.

Maps are written on the right.

All groups are finite, and all permutation representations are on finite sets.

D.C.C. denotes the *descending chain condition* or *Artinian condition* on right ideals of a ring, or on submodules of a module.

A.C.C. denotes the *ascending chain condition* or *Noetherian condition* on right ideals of a ring, or on submodules of a module.

$\square$ marks the end of a proof.

$\rightarrowtail$ denotes a monomorphism, and $\twoheadrightarrow$ denotes an epimorphism.

$\hookrightarrow$ denotes an inclusion.

$\mapsto$ denotes the action of a map on an element.

If H and K are subgroups of a group G, then $\sum\limits_{HgK}$ will denote a sum over a set of $H - K$ double coset representatives g in G.

$H \leq G$ means "H is a subgroup of G", and $H < G$ means "H is a proper subgroup of G". $N \trianglelefteq G$ means "N is a normal subgroup of G".

Note that the index is also an index of notation.

Section 1 Rings and Modules

1.1 The Jacobson Radical

Let Λ be a ring and V a Λ-module. The socle of V is the sum of all the irreducible submodules of V, and is written $Soc(V)$. The radical of V is the intersection of all the maximal submodules of V, and is written $Rad(V)$. V is said to be completely reducible if $V = Soc(V)$. The head of V is $V/Rad(V)$.

1.1.1 Lemma

If V satisfies D.C.C. then V is completely reducible if and only if $Rad(V) = 0$. In this case, V is a finite direct sum of irreducible modules.

The proof is an exercise. □

$J(\Lambda)$, the Jacobson radical of Λ, is defined to be the intersection of the annihilators of the irreducible Λ-modules, i.e. the intersection of the maximal right ideals of Λ. Let Λ_Λ denote Λ as a right Λ-module (called the regular representation of Λ). Then $J(\Lambda) = Rad(\Lambda_\Lambda)$. We say that Λ is semisimple if $J(\Lambda) = 0$. Note that $J(\Lambda/J(\Lambda))$ is always zero.

1.1.2 Lemma

Any right ideal of Λ is contained in a maximal right ideal of Λ.

Proof

Use Zorn's lemma and the fact that Λ has an identity element. □

1.1.3 Lemma

If $a \in J(\Lambda)$ then $1-a$ has a right inverse in Λ.

Proof

$1 = a + (1-a)$, and so $\Lambda = J(\Lambda) + (1-a)\Lambda$. If $(1-a)\Lambda \neq \Lambda$, choose M a maximal right ideal with $(1-a)\Lambda \subseteq M$ (1.1.2). Then $J(\Lambda) \subseteq M$ also, so $\Lambda \subseteq M$, a contradiction. □

1.1.4 Lemma (Nakayama)

If W is a finitely generated Λ-module and $W.J(\Lambda) = W$ then $W = 0$.

Proof

Suppose $W \neq 0$. Choose $w_1, \ldots, w_n$ generating W with n minimal. Since $W.J(\Lambda) = W$, we can write $w_n = \sum_{i=1}^{n} w_i a_i$ with $a_i \in J(\Lambda)$. By 1.1.3, $1-a_n$ has a right inverse b in Λ. Then $w_n(1-a_n) = \sum_{i=1}^{n-1} w_i a_i$, and so $w_n = (\sum_{i=1}^{n-1} w_i a_i).b$. This contradiction proves the lemma. □

1.1.5 Lemma

If Λ is semisimple and satisfies D.C.C. then every Λ-module is completely reducible.

Proof

$\mathrm{Rad}(\Lambda_\Lambda) = 0$, so by 1.1.1, Λ_Λ is completely reducible. Thus any module is a quotient of a direct sum of completely reducible modules, and is hence completely reducible. □

1.1.6 Theorem

If Λ satisfies D.C.C. then

(i) $J(\Lambda)$ is nilpotent.

(ii) If V is finitely generated then V satisfies A.C.C. and D.C.C. (i.e. V has a composition series).

(iii) Λ satisfies A.C.C.

Proof

(i) By D.C.C., for some n, $J(\Lambda)^n = J(\Lambda)^{2n}$, and if $J(\Lambda)^n \neq 0$, we may choose a minimal right ideal I with $IJ(\Lambda)^n \neq 0$. Choose $a \in I$ with $aJ(\Lambda)^n \neq 0$. Then $I = aJ(\Lambda)^n$ by minimality of I, and so for some $x \in J(\Lambda)^n$, $a = ax$. But then $a(1-x) = 0$, and so $a = 0$ by 1.1.3.

(ii) Let $V_i = V.J(\Lambda)^i$. Then V_i/V_{i+1} is annihilated by $J(\Lambda)$, and is hence completely reducible by 1.1.5. Since V is finitely generated it satisfies D.C.C. and hence so does V_i/V_{i+1}. Thus by 1.1.1, V_i/V_{i+1} has a composition series, and hence so does V.

(iii) follows by applying (ii) to Λ_Λ . □

1.2 The Wedderburn-Artin Structure Theorem

1.2.1 Lemma (Schur)

If V and W are simple Λ-modules, then for $V \not\cong W$, $\mathrm{Hom}_\Lambda(V,W) = 0$, while $\mathrm{Hom}_\Lambda(V,V) = \mathrm{End}_\Lambda(V) = E_\Lambda(V)$ is a division ring. □

An idempotent in Λ is a non-zero element e with $e^2 = e$.

1.2.2. Lemma

(i) If V is a Λ-module and e is an idempotent in Λ then $V.e \cong \mathrm{Hom}_\Lambda(e\Lambda,V)$.

(ii) $e\Lambda e \cong \mathrm{End}_\Lambda(e\Lambda)$.

Proof

(i) Define $f_1: Ve \to \mathrm{Hom}_\Lambda(e\Lambda, V)$ by $f_1(ve): ea \mapsto vea$ and $f_2: \mathrm{Hom}_\Lambda(e\Lambda, V) \to Ve$ by $f_2: \lambda \mapsto \lambda(e)$. Then f_1 and f_2 are inverse module homomorphisms.

(ii) follows from (i). □

1.2.3 Theorem

Let V be a completely reducible Λ-module with a composition series. Write $V = V_1 \oplus \ldots \oplus V_r$, with each V_i a direct sum of n_i modules $U_{i1} \oplus \ldots \oplus U_{in_i}$ isomorphic to a simple module U_i, and $U_i \not\cong U_j$ if $i \neq j$. Let $\Delta_i = \mathrm{End}_\Lambda(U_i)$. Then Δ_i is a division ring, $\mathrm{End}_\Lambda(V_i) \cong \mathrm{Mat}_{n_i}(\Delta_i)$, and $\mathrm{End}_\Lambda(V) = \bigoplus_i \mathrm{End}_\Lambda(V_i)$ is semisimple.

Proof

By 1.2.1, Δ_i is a division ring. Choose once and for all isomorphisms $\theta_{ij}: U_{ij} \to U_i$. Now given $\lambda \in \mathrm{End}_\Lambda(V_i)$, we define $\lambda_{jk} \in \Delta_i$ as the composite map

$$U_i \underset{\theta_{ij}}{\cong} U_{ij} \hookrightarrow V_i \xrightarrow[\lambda]{} V_i \twoheadrightarrow U_{ik} \underset{\theta_{ik}^{-1}}{\cong} U_i .$$

The map $\lambda \mapsto (\lambda_{jk})$ is then an injective homomorphism $\mathrm{End}_\Lambda(V_i) \to \mathrm{Mat}_{n_i}(\Delta_i)$. Conversely, given (λ_{jk}), we can construct λ as the sum of the composite endomorphisms

$$V_i \twoheadrightarrow U_{ij} \underset{\theta_{ij}^{-1}}{\cong} U_i \xrightarrow[\lambda_{jk}]{} U_i \underset{\theta_{ik}}{\cong} U_{ik} \hookrightarrow V_i .$$

Finally, $\mathrm{End}_\Lambda(V) = \bigoplus_i \mathrm{End}_\Lambda(V_i)$ since if $i \neq j$, 1.2.1 implies that $\mathrm{Hom}_\Lambda(V_i, V_j) = 0$. □

1.2.4 Theorem (Wedderburn-Artin Structure Theorem)

Let Λ be a semisimple ring satisfying D.C.C. Then $\Lambda = \bigoplus_{i=1}^{r} \Lambda_i$, $\Lambda_i \cong \mathrm{Mat}_{n_i}(\Delta_i)$, Δ_i is a division ring, and the Λ_i are uniquely determined. Λ has exactly r isomorphism classes of irreducible modules V_i with $\dim_{\Delta_i}(V_i) = n_i$. If Λ is simple then $\Lambda \cong \mathrm{Mat}_n(\Delta)$.

Proof

By 1.1.5, Λ_Λ is completely reducible. By 1.2.2 with $e = 1$, $\Lambda \cong \mathrm{End}_\Lambda(\Lambda_\Lambda)$. The result now follows by applying 1.2.3 to Λ_Λ. □

Remarks

(i) Wedderburn has shown that every division ring with finitely many elements is a field.

(ii) If Λ is a finite dimensional algebra over a field k, then each Δ_i for $\Lambda/J(\Lambda)$ has k in its centre. If for each i, we have $\Delta_i = k$, then k is called a <u>splitting field</u> for Λ.

If k_1 is an extension field of k, regarding $V \otimes_k k_1$ as a $\Gamma \otimes_k k_1$-module, we have $\mathrm{End}_{k_1}(V \otimes_k k_1) \cong \mathrm{End}_k(V) \otimes_k k_1$. Thus if k is a splitting field for Λ, every extension k_1 of k is a splitting field for $\Gamma \otimes_k k_1$. An algebraically closed field is a splitting field for any algebra over k of finite dimension.

<u>Exercises</u>

1. Suppose Λ is an n-dimensional semisimple commutative algebra over $\mathbb{C}$. Show that the number $f(n)$ of subalgebras of Λ (containing the identity element) satisfies $f(1) = 1$ and
$f(n+1) = f(n) + n.f(n-1) + \binom{n}{2}.f(n-2) + \ldots + n.f(1) + 1$.
Find $f(7)$.
2. Suppose k_1 and k_2 are finite extensions of k with k_1 separable. Show that $k_1 \otimes_k k_2$ is semisimple. Show that if Λ is a finite dimensional semisimple algebra over k and k_1 is a finite separable extension of k then $\Lambda \otimes_k k_1$ is semisimple.

1.3 <u>The Krull-Schmidt Theorem</u>

A ring R is said to be a <u>local ring</u> if the non-units in R form an ideal.

A Λ-module V is said to have the <u>unique decomposition property</u> if

(i) V is a direct sum of a finite number of indecomposable modules, and

(ii) Whenever $V = \bigoplus_{i=1}^{m} U_i = \bigoplus_{i=1}^{n} V_i$ with each U_i and each V_i ($\neq 0$) indecomposable, then $m = n$ and after reordering if necessary, $U_i \cong V_i$.

A ring Λ is said to have the unique decomposition property if every finitely generated Λ-module has.

1.3.1 <u>Theorem</u>

Suppose V is a finite sum of indecomposable Λ-modules V_i with the property that the endomorphism ring of each V_i is a local ring. Then V has the unique decomposition property.

<u>Proof</u>

Let $V = \bigoplus_{i=1}^{m} U_i = \bigoplus_{i=1}^{n} V_i$ and work by induction on n. Assume $n > 1$. Let

$$\alpha_i : U_i \hookrightarrow V \twoheadrightarrow V_1$$

and $$\beta_i : V_1 \hookrightarrow V \twoheadrightarrow U_i .$$

Then $1_{V_1} = \Sigma\beta_i\alpha_i : V_1 \to V_1$. Since $\mathrm{End}_\Lambda(V_1)$ is a local ring, some $\beta_i\alpha_i$ is a unit. Renumber so that $\beta_1\alpha_1$ is a unit. Then $U_1 \cong V_1$.

Consider the map $\mu = 1-\theta$, where

$$\theta : V \twoheadrightarrow V_1 \xrightarrow[\alpha_1^{-1}]{} U_1 \hookrightarrow V \twoheadrightarrow \bigoplus_{i=2}^{n} V_i \hookrightarrow V.$$

Then $U_1\mu = V_1$, and $(\bigoplus_{i=2}^{n} V_i)\mu = \bigoplus_{i=2}^{n} V_i$, so μ is onto. If $w\mu = 0$, then $w = w\theta$ and so $w \in \bigoplus_{i=2}^{n} V_i$. But then $w\theta = 0$.

Thus μ is an automorphism of V with $U_1\mu = V_1$, and so $\bigoplus_{i=2}^{n} U_i = V/U_1 \cong V/V_1 = \bigoplus_{i=2}^{n} V_i$. □

1.3.2 <u>Lemma</u> (Fitting)

Suppose V has a composition series and $f \in \mathrm{End}_\Lambda(V)$. Then for large enough n, $V = \mathrm{Im}(f^n) \oplus \mathrm{Ker}(f^n)$.

<u>Proof</u>

Choose n so that $f^n : Vf^n \to Vf^{2n}$ is an isomorphism. If $u \in V$, write $uf^n = vf^{2n}$. Then $u = vf^n + (u-vf^n) \in \mathrm{Im}(f^n) + \mathrm{Ker}(f^n)$. If $uf^n \in \mathrm{Im}(f^n) \cap \mathrm{Ker}(f^n)$ then $uf^{2n} = 0$, and so $uf^n = 0$. □

1.3.3 <u>Lemma</u>

Suppose V is indecomposable and has a composition series. Then $\mathrm{End}_\Lambda(V)$ is a local ring.

<u>Proof</u>

Let $E = \mathrm{End}_\Lambda(V)$, and choose I a maximal right ideal. Suppose $a \notin I$. Then $E = aE + I$. Write $1 = a\lambda + \mu$, $\lambda \in E$, $\mu \in I$. Since μ is not an isomorphism, 1.3.2 implies that $\mu^n = 0$ for some n. Thus $a\lambda(1+\mu+ \,..\, + \mu^{n-1}) = (1-\mu)(1+ \,..\, + \mu^{n-1}) = 1$, and so a is invertible. □

1.3.4 <u>Theorem</u> (Krull-Schmidt)

Suppose Λ satisfies D.C.C. Then Λ has the unique decomposition property.

<u>Proof</u>

Suppose V is a finitely generated indecomposable Λ-module. Then by 1.1.6 V has a composition series, and so by 1.3.3, $\mathrm{End}_\Lambda(V)$ is a local ring. The result now follows from 1.3.1. □

1.4 Cohomology of Modules

A module P is said to be projective if given two modules V and W and maps $\lambda: P \to V$ and $\mu: W \to V$ with μ epi, there is a map $\nu: P \to W$ such that

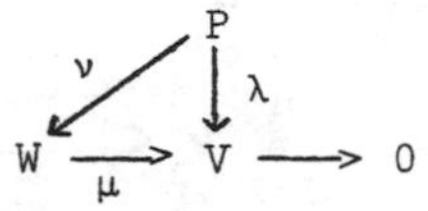

commutes.

A module I is injective if given two modules V and W and maps $\lambda: V \to I$ and $\mu: V \to W$ with μ mono, there is a map $\nu: W \to I$ such that

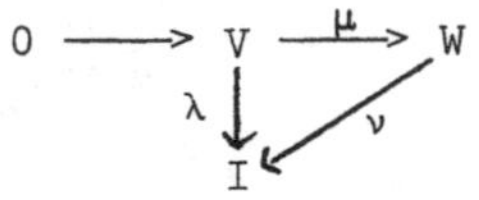

commutes.

1.4.1 Lemma

The following are equivalent:

(i) P is projective.

(ii) Every epimorphism $\lambda: V \to P$ splits.

(iii) P is a direct summand of a free module. □

1.4.2 Lemma

(i) Suppose $0 \longrightarrow W_1 \longrightarrow U_1 \xrightarrow{\sigma} V \longrightarrow 0$ and $0 \longrightarrow W_2 \longrightarrow P_2 \longrightarrow V \longrightarrow 0$ are short exact sequences, with P_2 projective, and suppose σ factors through a projective module. Then $W_1 \oplus P_2 \cong U_1 \oplus W_2$ **(This is needed in the proof of 2.27.1).**

(ii) (Schanuel's lemma) Suppose $0 \longrightarrow W_1 \longrightarrow P_1 \longrightarrow V \longrightarrow 0$ and $0 \longrightarrow W_2 \longrightarrow P_2 \longrightarrow V \longrightarrow 0$ are short exact sequences with P_1 and P_2 projective. Then $W_1 \oplus P_2 \cong P_1 \oplus W_2$.

Proof

(i) We construct a pullback diagram

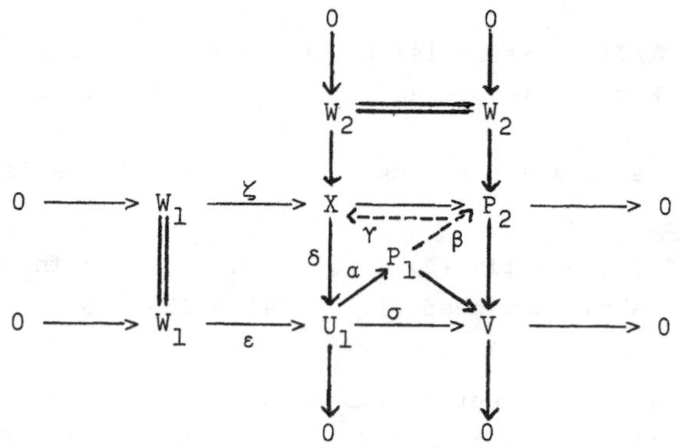

Since $(1-\alpha\beta\gamma\delta)\sigma = 0$, we may define

$$\theta = \alpha\beta\gamma + (1-\alpha\beta\gamma\delta)\varepsilon^{-1}\zeta.$$

Then θ is a splitting for δ , and so

$$W_1 \oplus P_2 \cong X \cong U_1 \oplus W_2.$$

(ii) This follows from (i). □

A map $\lambda\colon U \to V$ is called an <u>essential epimorphism</u> if no proper submodule of U is mapped onto V by λ , and it is called an <u>essential monomorphism</u> if every non-zero submodule of V has a non-zero intersection with $\mathrm{Im}(f)$. A <u>projective cover</u> for V is a projective module P together with an essential epimorphism $\lambda\colon P \to V$. An <u>injective hull</u> for V is an injective module I together with an essential monomorphism $\mu\colon V \to I$. The following theorem contains the necessary results about injective hulls and projective covers.

1.4.3. <u>Theorem</u>

(i) Projective covers, if they exist, are unique up to an isomorphism which commutes with the essential epimorphisms.

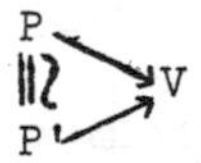

(ii) Injective hulls always exist, and are unique up to an isomorphism which commutes with the essential monomorphisms.

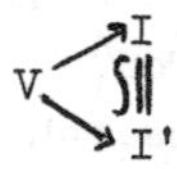

(iii) The following conditions on a ring Λ are equivalent :

(a) Every finitely generated Λ-module has a projective cover,

and (b) $\Lambda/J(\Lambda)$ satisfies D.C.C., and any decomposition of $\Lambda/J(\Lambda)$ as a direct sum of Λ-modules (c.f. 1.2.4) lifts to a decomposition of Λ_Λ.

(Note: a ring satisfying condition (b) is called semiperfect).

Proof.

(i), (ii) and (iii) (b) $\Rightarrow$ (a) can be found in Curtis and Reiner [38] §6. We shall not need (iii) (a) $\Rightarrow$ (b). □

Note

We shall see in section 1.5 that if Λ satisfies D.C.C. then Λ is semiperfect. In section 1.7 we shall see that a finitely generated algebra over a complete rank one discrete valuation ring is also semiperfect.

We write P_V for the projective cover of a module V, and I_V for its injective hull. We write $\Omega(V) = \mathrm{Ker}(P_V \to V)$ and $\Omega^{-1}(V) = \mho(V) = \mathrm{Coker}(V \to I_V)$.

Given a module V, a projective resolution of V is an infinite exact sequence

$\cdots \longrightarrow P_n \longrightarrow \cdots \longrightarrow P_2 \longrightarrow P_1 \longrightarrow V \longrightarrow 0$ with each P_i projective.

It is clear that every module has a projective resolution, since every module is a quotient of a free module. If W is another module, we get long sequences (not necessarily exact)

$$0 \longrightarrow \mathrm{Hom}_\Lambda(V,W) \xrightarrow[\delta_1]{} \mathrm{Hom}_\Lambda(P_1,W) \xrightarrow[\delta_2]{} \mathrm{Hom}_\Lambda(P_2,W) \xrightarrow[\delta_3]{} \cdots$$

and

$$\cdots \xrightarrow[\partial_3]{} P_2 \underset{\Lambda}{\otimes} W \xrightarrow[\partial_2]{} P_1 \underset{\Lambda}{\otimes} W \xrightarrow[\partial_1]{} V \underset{\Lambda}{\otimes} W \longrightarrow 0$$

with $\delta_i\delta_{i+1} = 0$ and $\partial_{i+1}\partial_i = 0$. We define

$$\mathrm{Ext}^i_\Lambda(V,W) = \mathrm{Ker}(\delta_{i+1})/\mathrm{Im}(\delta_i)$$

and

$$\mathrm{Tor}^\Lambda_i(V,W) = \mathrm{Ker}(\partial_i)/\mathrm{Im}(\partial_{i+1}) \qquad (i \geq 1).$$

The following facts are well known from homological algebra (see for example Cartan and Eilenberg 'Homological Algebra' or S. Maclane, 'Homological Algebra'; **see also section 2.22)**:

(i) The functors Ext^i_Λ and Tor^Λ_i are independent of the choice of projective resolution, in the sense that given two different projective resolutions of V, we get natural isomorphisms between the two functors Ext^i_Λ so defined, and likewise for Tor^Λ_i.

(ii) A short exact sequence $0 \to V_1 \to V_2 \to V_3 \to 0$ gives rise to long exact sequences

$$\begin{array}{l} 0 \longrightarrow \mathrm{Hom}_\Lambda(V_3,W) \longrightarrow \mathrm{Hom}_\Lambda(V_2,W) \longrightarrow \mathrm{Hom}_\Lambda(V_1,W) \\ \longrightarrow \mathrm{Ext}^1_\Lambda(V_3,W) \longrightarrow \mathrm{Ext}^1_\Lambda(V_2,W) \longrightarrow \mathrm{Ext}^1_\Lambda(V_1,W) \\ \longrightarrow \mathrm{Ext}^2_\Lambda(V_3,W) \longrightarrow \cdots \end{array}$$

and

$$\begin{array}{l} \cdots \longrightarrow \mathrm{Tor}^\Lambda_2(V_3,W) \\ \longrightarrow \mathrm{Tor}^\Lambda_1(V_1,W) \longrightarrow \mathrm{Tor}^\Lambda_1(V_2,W) \longrightarrow \mathrm{Tor}^\Lambda_1(V_3,W) \\ \longrightarrow V_1 \otimes_\Lambda W \longrightarrow V_2 \otimes_\Lambda W \longrightarrow V_3 \otimes_\Lambda W \longrightarrow 0. \end{array}$$

(iii) A short exact sequence $0 \to W_1 \to W_2 \to W_3 \to 0$ gives rise to a long exact sequence

$$\begin{array}{l} 0 \longrightarrow \mathrm{Hom}_\Lambda(V,W_1) \longrightarrow \mathrm{Hom}_\Lambda(V,W_2) \longrightarrow \mathrm{Hom}_\Lambda(V,W_3) \\ \longrightarrow \mathrm{Ext}^1_\Lambda(V,W_1) \longrightarrow \mathrm{Ext}^1_\Lambda(V,W_2) \longrightarrow \mathrm{Ext}^1_\Lambda(V,W_3) \\ \longrightarrow \mathrm{Ext}^2_\Lambda(V,W_1) \longrightarrow \cdots . \end{array}$$

(iv) The elements of $\mathrm{Ext}^1_\Lambda(V,W)$ may be interpreted as equivalence classes of short exact sequences $0 \to V \to X \to W \to 0$, two such being equivalent if there is a map of short exact sequences

$$\begin{array}{ccccccccc} 0 & \longrightarrow & V & \longrightarrow & X & \longrightarrow & W & \longrightarrow & 0 \\ & & \| & & \downarrow & & \| & & \\ 0 & \longrightarrow & V & \longrightarrow & X' & \longrightarrow & W & \longrightarrow & 0 \end{array} .$$

Note that by the five-lemma, the middle map $X \to X'$ is an isomorphism. See also [67] III.5 for a similar interpretation of Ext^i_Λ, $i > 1$.

We shall study cohomology for group algebras in more detail in section 2.22; what we have described here will suffice for our needs until then.

Now let Λ be an algebra over a field k. If V is a right Λ-module, then $V^* = \mathrm{Hom}_k(V,k)$ has a natural structure as a left Λ-module, and vice-versa. If V is finite dimensional as a vector space over k, then there is a natural isomorphism $(V^*)^* \cong V$. If V is injective, then V^* is projective, and vice-versa, since duality reverses all arrows.

In general, projective and injective modules for a ring are very different. However, there is a special situation under which they are the same. We say a finite dimensional algebra Λ over a field k is <u>Frobenius</u> if there is a linear map $\lambda : \Lambda \to k$ such that

(i) $\mathrm{Ker}(\lambda)$ contains no non-zero left or right ideal.

We say that Λ is <u>symmetric</u> if it satisfies (i) together with

(ii) For all $a, b \in \Lambda$, $\lambda(ab) = \lambda(ba)$.

1.4.4 <u>Proposition</u>

Let Λ be a Frobenius algebra over k. Then

(i) $({}_\Lambda\Lambda)^* \cong \Lambda_\Lambda$.

(ii) The following conditions on a finitely generated Λ-module are equivalent.

(a) V is projective

(b) V is injective

(c) V^* is projective

and (d) V^* is injective.

<u>Proof</u>

(i) We define a linear map $\varphi : \Lambda_\Lambda \longrightarrow ({}_\Lambda\Lambda)^*$ via $x\varphi : y \to \lambda(xy)$. Then if $\gamma \in \Lambda$, $y[(x\varphi)\gamma] = (\gamma y)(x\varphi) = \lambda(x\gamma y) = y[(x\gamma)\varphi]$, so φ is a homomorphism. By the defining property of λ, φ is injective, and hence surjective.

(ii) It follows from (i) that V is projective if and only if V^* is projective, so that (a) and (c) are equivalent. We have already remarked that (a) $\Leftrightarrow$ (d) and (b) $\Leftrightarrow$ (c) hold for all finite dimensional algebras. □

<u>Remarks</u>

(i) If Λ is the group algebra of a finite group (see 2.1), then Λ is symmetric, with $\lambda : \Sigma a_g.g \to a_1$. Then since $\lambda((\Sigma a_g.g).g^{-1}) = a_g$, property (i) is satisfied. Property (ii) is clear.

(ii) If Λ is Frobenius, and V has no projective direct summands, then $P_V \cong I_{\Omega(V)}$, and so $\mho(\Omega(V)) \cong V$. Similarly $\Omega(\mho(V)) \cong V$.

(iii) See the exercise to 1.5 for a property of symmetric algebras not shared by all Frobenius algebras.

<u>Exercises</u>

1. Let Δ be a finite dimensional division ring over its centre K. Show that $[\Delta,\Delta] = \text{span}(\{ab - ba \mid a, b \in \Delta\})$ is properly contained in Δ (Hint: let K_1 be a maximal subfield of Δ. Then $\Delta \otimes_K K_1 \cong \text{Mat}_n(K_1)$, and so $[\Delta \otimes_K K_1, \Delta \otimes_K K_1] \subsetneq \Delta \otimes_K K_1$). Deduce that every finite dimensional semisimple algebra over a field k is symmetric.

2. Show that if V and W are finitely generated $\mathbb{Z}$-modules, then $\text{Tor}_n^{\mathbb{Z}}(V,W) = 0$ for $n > 1$. Find $\text{Tor}_1^{\mathbb{Z}}(\mathbb{Z}/m\mathbb{Z}, \mathbb{Z}/n\mathbb{Z})$.

1.5 Idempotents and the Cartan Matrix

Recall that an idempotent in Λ is a non-zero element e with $e^2 = e$. If e is an idempotent than so is $1-e$. Two idempotents e_1 and e_2 are said to be orthogonal if $e_1e_2 = e_2e_1 = 0$. An idempotent e is primitive if we cannot write $e = e_1 + e_2$ with e_1 and e_2 orthogonal idempotents.

There is a one-one correspondence between expressions $1 = e_1 + .. + e_n$ with the e_i orthogonal idempotents, and direct sum decompositions $\Lambda_\Lambda = \Lambda_1 \oplus .. \oplus \Lambda_n$ of the regular representation, given by $\Lambda_i = e_i\Lambda$. Under this correspondence, e_i is primitive if and only if Λ_i is indecomposable.

1.5.1 Theorem (Idempotent Refinement)

Let N be a nilpotent ideal in Λ, and let e be an idempotent in Λ/N. Then there is an idempotent f in Λ with $e = \bar{f}$.

Proof

We define idempotents $e_i \in \Lambda/N^i$ inductively as follows. Let $e_1 = e$. For $i > 1$, let a be any element of Λ/N^i with image e_{i-1} in Λ/N^{i-1}. Then $a^2 - a \in N^{i-1}/N^i$, and so $(a^2-a)^2 = 0$. Let $e_i = 3a^2 - 2a^3$. Then e_i has image e_{i-1} in Λ/N^{i-1}, and

$$\begin{aligned} e_i^2 - e_i &= (3a^2-2a^3)(3a^2-2a^3-1) \\ &= -(3-2a)(1+2a)(a^2-a)^2 = 0. \end{aligned}$$

If $N^r = 0$, we take $f = e_r$. □

1.5.2 Corollary

Let N be a nilpotent ideal in Λ. Let $1 = e_1 + .. + e_n$ with the e_i primitive orthogonal idempotents in Λ/N. Then we can write $1 = f_1 + .. + f_n$ with the f_i primitive orthogonal idempotents in Λ and $\bar{f}_i = e_i$.

Proof

Define f_i' inductively as follows. $f_1' = 1$, and for $i > 1$, f_i' is any lift of $e_i + e_{i+1} + .. + e_n$ to an idempotent in the ring $f_{i-1}'\Lambda f_{i-1}'$. Then $f_i'f_{i+1}' = f_{i+1}' = f_{i+1}'f_i'$. Let $f_i = f_i' - f_{i+1}'$. Clearly $\bar{f}_i = e_i$. If $j > i$, $f_j = f_{i+1}'f_jf_{i+1}'$, and so $f_if_j = (f_i' - f_{i+1}')f_{i+1}'f_jf_{i+1}' = 0$. Similarly $f_jf_i = 0$. □

Now for the rest of section 1.5, suppose Λ satisfies D.C.C. Then by the Wedderburn-Artin structure theorem (1.2.4), we may write $\Lambda/J(\Lambda) = \bigoplus_{i=1}^{r} \Lambda_i$, $\Lambda_i \cong \mathrm{Mat}_{n_i}(\Delta_i)$. Let V_i be the irreducible

Λ-module corresponding to Δ_i, so that V_i is an n_i-dimensional module for Δ_i. Then $\Delta_i = V_{i1} \oplus .. \oplus V_{in_i}$ with $V_{ij} \cong V_i$. Corresponding to this decomposition of $\Lambda/J(\Lambda)$ we have a primitive orthogonal idempotent decomposition $1 = e_1 + .. + e_n$. By 1.5.2 we may lift this to a primitive orthogonal idempotent decomposition $1 = f_1 + .. + f_n$ in Λ, since $J(\Lambda)$ is nilpotent by 1.1.6. Letting $\Xi_i = f_i\Lambda$, we have $\Lambda_\Lambda = \Xi_1 \oplus .. \oplus \Xi_n$. By the Krull-Schmidt theorem (1.3.5) and 1.4.1, every projective indecomposable module is isomorphic to one of the Ξ_i. We say that the Ξ_i are the principal indecomposable modules for Λ (PIMs for short).

From this description, we see that $\Xi_i/\Xi_i J(\Lambda)$ is an irreducible module, and so Ξ_i has a unique maximal submodule. Moreover, the definition of a projective module ensures that any isomorphism $\Xi_i/\Xi_i J(\Lambda) \cong \Xi_j/\Xi_j J(\Lambda)$ lifts to a pair of maps $\Xi_i \to \Xi_j$ and $\Xi_j \to \Xi_i$. The composite is not nilpotent, so by Fitting's lemma (1.3.2) it is an isomorphism. Thus there are as many non-isomorphic PIMs as there are irreducible modules. If V_i are the irreducibles, write P_i for the corresponding PIMs. Thus we have shown that

$$\Lambda_\Lambda \cong \bigoplus_{i=1}^{r} n_i P_i \quad .$$

In particular, we have shown that if Λ satisfies D.C.C. then Λ is semiperfect, and so every module has a projective cover (see 1.4.3), namely the unique projective module with isomorphic head.

We say that two primitive idempotents e and e' in Λ are equivalent if they lie in the same Wedderburn component of $\Lambda/J(\Lambda)$, or equivalently if $e\Lambda$ and $e'\Lambda$ are isomorphic PIMs.

1.5.3 Lemma

$$\mathrm{Hom}_\Lambda(P_i,V_j) = \begin{cases} \Delta_i & \text{if } i = j \\ 0 & \text{otherwise} \end{cases} \quad .$$

Proof

Every homomorphism from P_i to a simple module factors through V_i. □

1.5.4 Lemma

$\mathrm{Dim}_{\Delta_i} \mathrm{Hom}_\Lambda(P_i,V)$ is the multiplicity of V_i as a composition factor of V.

Proof

Use 1.5.3 and induction on the composition length of V. An exact sequence $0 \to V' \to V \to V_j \to 0$ induces an exact sequence $0 \to \mathrm{Hom}_\Lambda(P_i,V') \to \mathrm{Hom}_\Lambda(P_i,V) \to \mathrm{Hom}_\Lambda(P_i,V_j) \to 0$. □

The Cartan invariants of Λ are defined as

$$\begin{aligned} c_{ij} &= \dim_{\Delta_i} \mathrm{Hom}_\Lambda(P_i,P_j) \\ &= \text{multiplicity of } V_i \text{ as a composition factor of } P_j. \end{aligned}$$

The matrix c_{ij} is called the Cartan matrix of the ring Λ. In general, the matrix c_{ij} may be singular (e.g. $\Lambda = \mathbb{F}_3S_3/J(\mathbb{F}_3S_3)^2$). We shall see in section 2.11, however, that this never happens for group algebras of finite groups. In fact, in 2.16.5 we shall show that the determinant of the Cartan matrix of a group algebra over a field of characteristic $p > 0$ is a power of p.

Finally, we shall need the following result.

1.5.5 Lemma (Rosenberg)

Suppose e is a primitive idempotent in a ring Λ satisfying D.C.C., and $e \in I_1 + .. + I_r$ with the I_j two-sided ideals. Then for some j, $e \in I_j$.

Proof

By 1.2.2, $e\Lambda e \cong \mathrm{End}_\Lambda(e\Lambda)$. By 1.1.6 and 1.3.3 this is a local ring. Each eI_je is an ideal in $e\Lambda e$, and so for some j, $e \in eI_je \subseteq I_j$. □

Exercise

Suppose Λ is a symmetric algebra. Let e be a primitive idempotent in Λ and let $P = e\Lambda$ be the corresponding PIM. Thus $V = P/\mathrm{Rad}(P)$ and $W = \mathrm{Soc}(P)$ are simple. Use the map λ to show that $We \neq 0$. Show that there is a non-zero homomorphism from P to W. Deduce that $V \cong W$.

Using the dual of lemma 1.5.4, show that if Λ is a symmetric algebra and k is a splitting field, then the Cartan matrix for Λ is symmetric.

1.6 Blocks and Central Idempotents

A central idempotent in Λ is an idempotent in the centre of Λ. A primitive central idempotent is a central idempotent not expressible as the sum of two orthogonal central idempotents. There is a one-one correspondence between expressions $1 = e_1 + .. + e_s$ with e_i orthogonal central idempotents and direct sum decompositions $\Lambda = B_1 \oplus .. \oplus B_s$ of Λ as two-sided ideals, given by $B_i = \Lambda e_i$.

Now suppose Λ satisfies D.C.C. Then we can write

$\Lambda = B_1 \oplus .. \oplus B_s$ with the B_i indecomposable as two-sided ideals.

1.6.1 Lemma

This decomposition is unique; i.e. if $\Lambda = B_1 \oplus .. \oplus B_s = B_1' \oplus .. \oplus B_t'$ then $s = t$ and for some permutation σ of $\{1, .. , s\}$, $B_i = B_{\sigma(i)}'$.

Proof

Write $1 = e_1 + .. + e_s = e_1' + .. + e_t'$. Then $e_i e_j'$ is also a central idempotent (or zero) for each i, j. Thus $e_i = e_i e_1' + .. + e_i e_t'$, so that for a unique j, $e_i = e_i e_j' = e_j'$. □

The indecomposable two-sided ideals in this decomposition are called the blocks of Λ.

Suppose V is indecomposable. Then $V = Ve_1 \oplus .. \oplus Ve_s$ shows that for some i, $Ve_i = V$, and $Ve_j = 0$ for $j \neq i$. We then say that V belongs to the block B_i. Thus the simples and PIMs are classified into blocks. Clearly if a module is in a certain block, then so are all its composition factors. Thus if V_i and V_j are in different blocks, then $c_{ij} = 0$.

The central primitive idempotents are thus also called the block idempotents.

If Λ is a finite dimensional algebra over a splitting field k, the algebra homomorphisms ω: $Z(\Lambda) \to k$ are called the central homomorphisms.

1.6.2 Proposition

(i) Let $Z(\Lambda)$ be the centre of Λ. Then

$$Z(\Lambda) = Z(\Lambda)e_1 \oplus .. \oplus Z(\Lambda)e_s$$

is the block decomposition of $Z(\Lambda)$. Each $Z(\Lambda)e_i$ is a local ring, and we have an inclusion map

$$Z(\Lambda)e_i/J(Z(\Lambda)e_i) \hookrightarrow \mathrm{End}_\Lambda(V)$$

for each irreducible Λ-module V in B_i.

(ii) Suppose Λ is a finite dimensional algebra over a splitting field k. Then k is also a splitting field for $Z(\Lambda)$, and in particular

$$Z(\Lambda)e_i/J(Z(\Lambda)e_i) \cong k.$$

There is a one-one correspondence between central homomorphisms ω_i and blocks B_i, with the property that $\omega_i(e_j) = \delta_{ij}$.

Proof

(i) A decomposition of 1 as a sum of central idempotents in Λ

and in $Z(\Lambda)$ are the same thing, so that

$$Z(\Lambda) = Z(\Lambda)e_1 \oplus .. \oplus Z(\Lambda)e_s$$

is the block decomposition of $Z(\Lambda)$. By 1.2.2(ii) and 1.3.3, since $Z(\Lambda)$ is commutative, we have that $Z(\Lambda)e_i \cong \mathrm{End}_{Z(\Lambda)}(Z(\Lambda)e_i)$ is a local ring.

If V is an irreducible Λ-module in B_i, then $Z(\Lambda)e_i$ acts non-trivially on V as endomorphisms, since e_i acts as the identity element. Thus we have a non-trivial map

$$Z(\Lambda)e_i \to \mathrm{End}_\Lambda(V).$$

Since $Z(\Lambda)e_i$ is a local ring and $\mathrm{End}_\Lambda(V)$ is a division ring, this induces an injection

$$Z(\Lambda)e_i/J(Z(\Lambda)e_i) \hookrightarrow \mathrm{End}_\Lambda(V).$$

(ii) If k is a splitting field for Λ, then we have maps

$$k \hookrightarrow Z(\Lambda)e_i/J(Z(\Lambda)e_i) \hookrightarrow \mathrm{End}_\Lambda(V) = k$$

whose composite is the identity. Thus $Z(\Lambda)/J(Z(\Lambda))$ is a direct sum of s copies of k, and the central homomorphisms ω_i are simply the s projection maps. Hence $\omega_i(e_j) = \delta_{ij}$. □

Exercise.

Show that every commutative algebra satisfying D.C.C. is a direct sum of local rings.

1.7 Algebras over a Valuation Ring

Let R be a complete rank one discrete valuation ring in characteristic zero (e.g. a p-adic completion of an algebraic number ring) and let (π) be its maximal ideal. Let $\hat{R}$ denote the field of fractions of R, and let $\bar{R} = R/(\pi)$ be a field of characteristic $p \neq 0$. We then say that $(\hat{R},R,\bar{R})$ is a p-modular system. Let Λ be an algebra over R, which as an R-module is free and of finite rank. Let $\hat{\Lambda} = \Lambda \otimes_R \hat{R}$ and $\bar{\Lambda} = \Lambda \otimes_R \bar{R} = \Lambda/\Lambda(\pi)$, and suppose $\hat{\Lambda}$ is semi-simple. From now on, when we talk of Λ-modules, we shall mean finitely generated R-free Λ-modules. Similarly, we only consider finitely generated $\hat{\Lambda}$-modules and $\bar{\Lambda}$-modules, which are respectively called ordinary and modular representations. If V is a Λ-module, we let $\hat{V} = V \otimes_R \hat{R}$ as a $\hat{\Lambda}$-module, and $\bar{V} = V \otimes_R \bar{R} = V/V(\pi)$ as a $\bar{\Lambda}$-module. If $\hat{R}$ is a splitting field for $\hat{\Lambda}$ and $\bar{R}$ is a splitting field for $\bar{\Lambda}$, we say that $(\hat{R},R,\bar{R})$ is a splitting p-modular system for Λ .

1.7.1 Lemma

If W is a $\hat{\Lambda}$-module, there is a Λ-module X with $\hat{X} \cong W$.

Proof

Choose a basis $w_1, .. , w_n$ for W over $\hat{R}$ and let $X = w_1\Lambda + .. + w_n \Lambda$. X is torsion free, and hence free. Choose a free basis $x_1, .. , x_m$. Then the x_i span W and are $\hat{R}$ independent, and hence $m = n$, and $W = X \otimes_R \hat{R}$. □

Such a Λ-module X is called an R-form of W. In general, R-forms are not unique.

1.7.2 Theorem (Idempotent Refinement)

(i) Let e be an idempotent in $\bar{\Lambda}$. Then there is an idempotent f in Λ with $e = \bar{f}$.

(ii) Let $1 = e_1 + .. + e_n$ with the e_i primitive orthogonal idempotents in $\bar{\Lambda}$. Then we can write $1 = f_1 + .. + f_n$ with the f_i primitive orthogonal idempotents in Λ, and $\bar{f}_i = e_i$.

(iii) Let $1 = e_1 + .. + e_s$ with the e_i primitive central idempotents in $\bar{\Lambda}$. Then we can write $1 = f_1 + .. + f_s$ with the f_i primitive central idempotents in Λ, and $\bar{f}_i = e_i$.

Proof

(i) We may apply 1.5.1 to $\Lambda/\Lambda(\pi^n)$ to obtain elements $f_i \in \Lambda$ whose image in $\Lambda/\Lambda(\pi^i)$ is a lift of f_{i-1} to an idempotent. Then the f_i form a Cauchy sequence, and we can take f as their limit.

(ii) Apply the same argument to 1.5.2.

(iii) Apply (ii) to the centre of Λ . □

Thus the decomposition of $\bar{\Lambda}_{\bar{\Lambda}}$ as a sum of PIMs lifts to a decomposition of Λ_Λ. So given an irreducible $\bar{\Lambda}$-module V_j, it has a projective cover $P_j = \bar{Q}_j$ for some projective Λ-module Q_j. We define the decomposition numbers d_{ij} as follows. Let $W_1, .. , W_t$ be the irreducible $\hat{\Lambda}$-modules, $X_1, .. , X_t$ be R-forms of them, and $V_1, .. , V_r$ be the irreducible $\bar{\Lambda}$-modules. We define

$\hat{\mu}_i = \dim_{\hat{R}} \mathrm{End}_{\hat{\Lambda}}(W_i)$

$\bar{\mu}_i = \dim_{\bar{R}} \mathrm{End}_{\bar{\Lambda}}(V_i)$

$d_{ij}/\hat{\mu}_i$ = multiplicity of W_i as a summand of $\hat{Q}_j$.

Remark

Note that when $(\hat{R},R,\bar{R})$ is a splitting p-modular system, all the $\hat{\mu}_i$ and $\bar{\mu}_i$ are one.

Then $d_{ij} = \dim_{\hat{R}}\mathrm{Hom}_{\hat{\Lambda}}(\hat{Q}_j, W_i)$ (see exercise 3)

$= \mathrm{rank}_R \mathrm{Hom}_\Lambda(Q_j, X_i)$ since $\mathrm{Hom}_\Lambda(Q_j, X_i)$ is an R-form for $\mathrm{Hom}_{\hat{\Lambda}}(\hat{Q}_j, W_i)$

$= \dim_{\bar{R}}\mathrm{Hom}_{\bar{\Lambda}}(P_j, \bar{X}_i)$ since Q_j is projective, and so

$d_{ij}/\bar{\mu}_j$ = multiplicity of V_j as a composition factor of $\bar{X}_i$.

In particular

$$c_{ij} = \sum_k d_{ki}d_{kj}/\hat{\mu}_k\bar{\mu}_i$$

and if $(\hat{R},R,\bar{R})$ is a splitting system then the Cartan matrix is symmetric.

Note carefully what the above is saying in case $(\hat{R},R,\bar{R})$ is a splitting system. It is saying that the decomposition matrix (d_{ij}) may be read two different ways up. The columns give the ordinary composition factors of a projective indecomposable (tensored with $\hat{R}$), and the rows give the modular composition factors of the reduction modulo (π) of an R-form of the ordinary irreducible. Exercise 3 shows that the decomposition numbers are well-defined, and so in particular the modular composition factors of an ordinary irreducible do not depend upon the R-form chosen.

By 1.7.2(iii), there is a one-one correspondence between blocks of Λ and blocks of $\bar{\Lambda}$, having the property that if V is an indecomposable Λ-module, then all summands of $\bar{V}$ are in the block corresponding to V. We shall identify corresponding blocks, so that we regard all indecomposable $\hat{\Lambda}$, Λ and $\bar{\Lambda}$-modules as falling into blocks of Λ. It is clear that if W_i is in a different block from V_j then $d_{ij} = 0$.

If $(\hat{R},R,\bar{R})$ is a splitting p-modular system for Λ, then given an irreducible $\hat{\Lambda}$-module W_i with central homomorphism $\omega_i : Z(\hat{\Lambda}) \to \hat{R}$, $\omega_i(Z(\Lambda)) \subseteq R$ (since any subring of $\hat{R}$ which is finitely generated as an R-module is contained in R), and so we get a central homomorphism $\bar{\omega}_i : Z(\bar{\Lambda}) \to \bar{R}$. This central homomorphism determines the block to which W_i belongs.

Exercises

1. (i) Let V be a Λ-module and $f \in \mathrm{End}_\Lambda(V)$. Write $\mathrm{Im}(f^\infty) = \bigcap_{n=1}^{\infty} \mathrm{Im}(f^n)$ and $\mathrm{Ker}(f^\infty) = \{x \in V : \forall\, n \geq 0\ \exists\, m \geq 0 \text{ s.t. } xf^m \in V.J(\Lambda)^n\}$.
Using Fitting's lemma (1.3.2) show that $V = \mathrm{Im}(f^\infty) \oplus \mathrm{Ker}(f^\infty)$.

(ii) Modify the argument of 1.3.3 to show that if V is an

indecomposable Λ-module then $End_\Lambda(V)$ is a local ring.

(iii) Apply 1.3.1 to deduce that Λ has the unique decomposition property.

2. Check that lemma 1.5.5 holds for primitive idempotents in Λ.

3. If P is a PIM for $\bar{\Lambda}$, show that any two lifts Q_1 and Q_2 of P to a Λ-module are isomorphic (Hint: use exercise 1). Deduce that the decomposition numbers are well defined.

1.8 A Little Commutative Algebra

We shall occasionally need a little commutative algebra, and so I have collected here, for easy reference, all the results that will be used.

The following is a version of the 'going-up' theorem, and will be needed in section 2.9.

1.8.1 Proposition

Suppose A is a commutative algebra over an algebraically closed field k, and B is integral over A. If $\lambda: A \to k$ is an algebra homomorphism, then there exists an algebra homomorphism $\mu: B \to k$ such that the restriction of μ to A is equal to λ (μ is not necessarily unique).

Proof

We extend λ a bit at a time. Choose $b \in B\backslash A$, and let the minimal equation of b be $b^n + a_{n-1}b^{n-1} + .. + a_o = 0$, $a_i \in A$. Since k is algebraically closed, the equation $x^n + (a_{n-1}\lambda)x^{n-1} + .. + (a_o\lambda) = 0$ has a solution $x = \zeta$ in k. We may then send $\langle A, b \rangle$ to k by sending b to ζ. By Zorn's lemma, we may continue until we have a homomorphism μ extending λ. □

When we come to study Poincaré series in sections 2.22 and 2.25 (see the paragraph before 2.25.17), we shall need the following proposition, whose proof I have lifted from Atiyah and Macdonald, 'Commutative Algebra'.

1.8.2 Proposition

Suppose A is a commutative graded Noetherian ring over a field k, with each A_i finite dimensional over k, and M is a finitely generated graded A-module. Then the Poincaré series $P(M,t) = \sum_i t^i \dim_k(M_i)$ is of the form $p(t)/\prod_{j=1}^{s}(1-t^{k_j})$, where $p(t)$ is a polynomial with integer coefficients and $k_1, .. , k_s$ are the degrees of a set of homogeneous generators of A over A_o.

Proof

Suppose A is generated over A_o by homogeneous elements $x_1, \ldots, x_s$ of degrees $k_1, \ldots, k_s$, and work by induction on s. If $s = 0$, then $\dim(M_n) = 0$ for all large n, and so $P(M,t)$ is a polynomial with integer coefficients.

Now suppose $s > 0$. Multiplication by x_s gives an exact sequence

$$0 \longrightarrow K_n \longrightarrow M_n \xrightarrow{x_s} M_{n+k_s} \longrightarrow L_{n+k_s} \longrightarrow 0,$$

where K and L are the kernel and cokernel of multiplication by x_s, and are hence finitely generated graded $A_o[x_1, \ldots, x_{s-1}]$-modules. Taking Poincaré series, we obtain

$$(1-t^{k_s})P(M,t) = P(L,t) - t^{k_s}P(K,t) + g(t)$$

where $g(t)$ is a polynomial with integer coefficients. The result now follows from the inductive hypothesis. □

Remark

This result also holds, without change in the proof, if A is 'graded commutative', in the sense that for homogeneous elements x and y, we have

$$xy = (-1)^{\deg(x)\deg(y)}\, yx.$$

This will be the form in which we shall use the proposition.

Suppose A is a commutative graded Noetherian ring and M is a finitely generated graded A-module. Then we write $\gamma(M)$ for the degree of the pole of $P(M,t)$ at $t = 1$. The following proposition relates $\gamma(M)$ to the rate of growth of the coefficients $\dim(M_i)$ of $P(M,t)$. **This will be used in section 2.31.**

1.8.3 Proposition

Let $f(t)$ be a rational function of the form $p(t)/\prod_{j=1}^{s}(1-t^{k_j})$ $= \sum_{i=0}^{\infty} a_i t^i$, where the a_i **are non-negative** integers. Let c be the order of the pole of $f(t)$ at $t = 1$. Then

(i) There is a positive number λ such that for all large enough n, $a_n \le \lambda n^{c-1}$, but there is no positive number μ such that $a_n \le \mu n^{c-2}$ for all large enough n.

(ii) The value of the analytic function $(\prod_i k_i).f(t)(1-t)^c$ at $t = 1$ is a positive integer.

Proof

The hypotheses and conclusion remain unaltered if we replace $f(t)$ by $f(t)(1+t+ ..+t^{k_j-1})$, and so without loss of generality we may assume that each k_j is one. We may thus assume that $f(t) = p(t)/(1-t)^c$, where $p(t) = \alpha_m t^m + .. + \alpha_o$ satisfies $p(1) \neq 0$, i.e. $\alpha_m + .. + \alpha_o \neq 0$. Thus

$$a_n = \alpha_o \binom{n+c-1}{c-1} + \alpha_1 \binom{n+c-2}{c-1} + .. + \alpha_m \binom{n+c-m+1}{c-1}$$

is a polynomial in n of degree $c-1$, thus proving (i). If $p(1)$ were negative, then for large n, a_n would also be negative, and so (ii) is proved. □

Finally, we shall need 1.8.6 in the proof of 2.24.4(xi).

1.8.4 Lemma

Let A be a commutative graded ring, I a homogeneous ideal, and $P^{(1)}, .., P^{(r)}$ homogeneous prime ideals. If $I \subsetneq P^{(i)}$ for each $1 \le i \le r$, then there is a homogeneous element in I which is not in any of the $P^{(i)}$.

Proof

We shall proceed by induction on r. The case $r = 1$ is clear. Suppose first that $r = 2$. Suppose every homogeneous element of I is in $P^{(1)} \cup P^{(2)}$. Choose a homogeneous element x of degree j in $I \backslash P^{(1)}$ and y of degree k in $I \backslash P^{(2)}$. Then $x \in P^{(2)}$ and $y \in P^{(1)}$, and so $x^k + y^j$ is a homogeneous element of I which is not in $P^{(1)}$ or $P^{(2)}$.

Now suppose $r > 2$. If $P^{(i)} \subseteq P^{(r)}$ for some $i < r$, we may delete $P^{(i)}$ and the result follows by induction, so we may assume $P^{(i)} \not\subseteq P^{(r)}$ for $i < r$. Hence $IP^{(1)}..P^{(r-1)} \not\subseteq P^{(r)}$, so by the result for $r = 1$, there is a homogeneous element x of degree j in $IP^{(1)}..P^{(r-1)} \backslash P^{(r)}$. By the inductive hypothesis, there is a homogeneous element y of degree k in $I \backslash (P^{(1)} .. P^{(r-1)})$. Suppose every homogeneous element of I is in $P^{(1)} \cup .. \cup P^{(r)}$. Then $y \in P^{(r)}$, and so $x^k + y^j$ is a homogeneous element which is in I but not in any of the $P^{(i)}$. □

Notation

If M is a graded module for a graded ring A, we write $M(r)$ for the 'twisted' module with $M(r)_i = M_{i+r}$, and the same A-action.

1.8.5 Lemma

Let A be a commutative graded Noetherian ring and M a

finitely generated graded A-module. Then M has a filtration

$$M = M^{(t)} \supset M^{(t-1)} \supset .. \supset M^{(0)} = 0,$$

with $M^{(i)}/M^{(i-1)} \cong (A/P^{(i)})(r_i)$ and $P^{(i)}$ (not necessarily distinct) homogeneous prime ideals.

Proof

Let $P^{(1)}$ be a maximal element of the set of annihilators of non-zero homogeneous elements $x \in M$. Then $P^{(1)}$ is a homogeneous ideal. If a and b are homogeneous, $ab \in P^{(1)}$, and $a \notin P^{(1)}$, then $xa \neq 0$ and $xab = 0$. Since $P^{(1)} = \mathrm{Ann}(x) \subseteq \mathrm{Ann}(xa)$, maximality of $P^{(1)}$ implies that $P^{(1)} = \mathrm{Ann}(xa)$, and so $b \in P^{(1)}$. Thus $P^{(1)}$ is prime. If $\deg(x) = r_1$, then the map $a \mapsto xa$ is an injection of $(A/P^{(1)})(r_1)$ into M. Let its image be $M^{(1)}$. Proceeding in the same way for $M/M^{(1)}$, we obtain an injection of $(A/P^{(2)})(r_2)$ into $M/M^{(1)}$. Let $M^{(2)}/M^{(1)}$ be its image. Continuing in this way, we obtain an ascending chain of submodules $0 = M^{(0)} \subset M^{(1)} \subset M^{(2)} \subset \ldots$. Since M satisfies A.C.C., for some t we have $M^{(t)} = M$. □

1.8.6 Proposition

Let A be a commutative graded Noetherian ring, and M a finitely generated graded A-module. Then there is a homogeneous element x of positive degree j in A such that for all n sufficiently large, multiplication by x induces an injective map $M_n \to M_{n+j}$.

Proof

By 1.8.5, M has a filtration $M = M^{(t)} \supset M^{(t-1)} \supset .. \supset M^{(0)} = 0$, with $M^{(i)}/M^{(i-1)} \cong (A/P^{(i)})(r_i)$ and $P^{(i)}$ homogeneous prime ideals. Denote by I the ideal of A generated by the elements of positive degree. By 1.8.4, we may choose an element x of degree j which lies in I, but does not lie in any of those $P^{(i)}$ not containing I. For $n > \max(r_i)$, suppose $u \in M_n$. Suppose u lies in $M^{(i)}$ but not in $M^{(i-1)}$. Then u has non-zero image $\bar{u}$ in $M^{(i)}/M^{(i-1)} \cong (A/P^{(i)})(r_i)$, and since $n > r_i$, this implies that $P^{(i)} \not\supseteq I$, and so $\bar{u}x \neq 0$. Hence $ux \neq 0$, and so multiplication by x induces an injection $M_n \to M_{n+j}$. □

1.8.7 Proposition

Let A be a commutative graded Noetherian ring with $A_0 = k$, and M a finitely generated graded A-module. Then $\gamma(M)$ is equal to the Krull dimension of $A/\mathrm{ann}(M)$, i.e. the maximal length n of a chain of homogeneous prime ideals

$$A \supset P^{(o)} \supset P^{(1)} \supset .. \supset P^{(n)} \supseteq \mathrm{ann}(M).$$

(geometrically, this is the dimension of Spec(A/ann(M)) as a variety; namely one more than the dimension of Proj(A/ann(M))).

Proof

By 1.8.5, we may assume that $M = A/P$ with P prime, and without loss of generality $P = 0$.

We first show that $\dim(A) \le \gamma(A)$, by induction on $\gamma(A)$. If $\gamma(A) = 0$, then $P(A,t)$ is a polynomial, and so A is finite dimensional over k. Thus all elements of positive degree are nilpotent, and hence form the only prime ideal (which is hence zero). Now suppose $\gamma(A) > 0$, and suppose $A \supset P^{(o)} \supset \ldots \supset P^{(n)} = 0$ is a chain of prime ideals. Let x be a homogeneous element of degree d in $P^{(n-1)}$, and let (x) be the principal ideal of A generated by x. Then by the inductive hypothesis, $n-1 \le \gamma(A/(x))$. But x is not a zero divisor, and so the exact sequence

$0 \longrightarrow A \xrightarrow{x} A \longrightarrow A/(x) \longrightarrow 0$ shows that $P(A/(x),t) = (1-t^d)P(A,t)$. Thus $\gamma(A/(x)) = \gamma(A) - 1$.

Conversely, we now show that $\gamma(A) \le \dim(A)$. This time we proceed by induction on $\dim(A)$, which we may since we now know that it is finite. If $\dim(A) = 0$, the only prime ideal is the one consisting of the elements of positive degree. But zero is a prime ideal, and so A is one dimensional, and $\gamma(A) = 0$. Now suppose $\dim(A) > 0$. Let x be a non-zero homogeneous element of positive degree. As before, we have $\gamma(A/(x)) = \gamma(A) - 1$. By the inductive hypothesis, there is a chain $A \supset P^{(o)} \supset .. \supset P^{(n-1)} \supseteq A/(x)$, where $n = \gamma(A)$. Thus $A \supset P^{(o)} \supset .. \supset P^{(n-1)} \supset P^{(n)} = 0$ is a chain of length n in A, and so $\dim(A) \ge n$. □

Section 2. Modules for Group Algebras

Throughout section 2, G will denote a finite group, $(\hat{R},R,\bar{R})$ will be a p-modular system (see section 1.7), k will be an arbitrary field of characteristic p, and Γ will be an arbitrary commutative ring with the property that ΓG has the unique decomposition property (see section 1.3); e.g. $\Gamma \in \{\hat{R},R,\bar{R}\}$.

2.1 Tensors and Homs; Induction and Restriction

We define the group algebra ΓG to be the free Γ-module on the elements of G, with multiplication defined as the linear extension of the multiplication in G.

Maschke's theorem [51, p. 91] shows that $\hat{R}G$ is semisimple, and so the theory of 1.7 gives us decomposition numbers and Cartan invariants, relating the projective indecomposables, the ordinary (i.e. characteristic 0) irreducibles and the modular (i.e. characteristic p) irreducibles.

Note that kG is a symmetric algebra (see remark after 1.4.4). Thus projective modules are the same thing as injective modules, and the unique irreducible submodule of a projective indecomposable is isomorphic to its unique irreducible quotient (1.4.4 and the exercise to 1.5).

Recall our convention (see 1.7) that all ΓG-modules are finitely generated and Γ-free. If V and W are ΓG-modules, we define ΓG-module structures on $V \underset{\Gamma}{\otimes} W$ and $\mathrm{Hom}_\Gamma(V,W)$ in the usual way:

$$(v \otimes w)g = vg \otimes wg$$

$$v(\lambda g) = ((vg^{-1})\lambda)g$$

$$(v \in V,\ w \in W,\ g \in G,\ \lambda \in \mathrm{Hom}_\Gamma(V,W)).$$

We let Γ be the trivial ΓG-module, and define $V^* = \mathrm{Hom}_\Gamma(V,\Gamma)$. The following facts are elementary.

2.1.1 Lemma

There are natural ΓG-module isomorphisms

(i) $V \otimes W \cong W \otimes V$

(ii) $V \otimes (W_1 \oplus W_2) \cong V \otimes W_1 \oplus V \otimes W_2$

(iii) $(U \otimes V) \otimes W \cong U \otimes (V \otimes W)$

(iv) $V^* \otimes W \cong \mathrm{Hom}_\Gamma(V,W)$

(this is also an $\mathrm{End}_{\Gamma G}(V)$ - $\mathrm{End}_{\Gamma G}(W)$ bimodule isomorphism)

(v) $(V^*)^* \cong V$. □

If H is a subgroup of G, and V is a ΓG-module, we write

$V{\downarrow}_H$ for the restriction of V as a ΓH-module. If W is a ΓH-module, we write $W{\uparrow}^G$ for the induced module $W \underset{\Gamma H}{\otimes} \Gamma G$. The following lemmas contain the elementary facts on restriction and induction.

2.1.2 Lemma

Let V, V_1 and V_2 be ΓG-modules and W, W_1 and W_2 be ΓH-modules. There are natural isomorphisms

(i) $(V_1 \oplus V_2){\downarrow}_H \cong V_1{\downarrow}_H \oplus V_2{\downarrow}_H$

(ii) $(V_1 \otimes V_2){\downarrow}_H \cong V_1{\downarrow}_H \otimes V_2{\downarrow}_H$

(iii) $(W_1 \oplus W_2){\uparrow}^G \cong W_1{\uparrow}^G \oplus W_2{\uparrow}^G$

(iv) $V \otimes W{\uparrow}^G \cong (V{\downarrow}_H \otimes W){\uparrow}^G$

(v) $(\mathrm{Hom}_\Gamma(V{\downarrow}_H, W)){\uparrow}^G \cong \mathrm{Hom}_\Gamma(V, W{\uparrow}^G)$

(vi) $(\mathrm{Hom}_\Gamma(W, V{\downarrow}_H)){\uparrow}^G \cong \mathrm{Hom}_\Gamma(W{\uparrow}^G, V)$. □

We write $(V,W)^G$ for $\mathrm{Hom}_{\Gamma G}(V,W)$ and W^G for $\mathrm{Hom}_{\Gamma G}(\Gamma, W)$, the set of fixed points of G on W.

2.1.3 Lemma (Frobenius Reciprocity)

There are natural isomorphisms

(i) $Fr_{H,G} : (V{\downarrow}_H, W)^H \cong (V, W{\uparrow}^G)^G$
given by $v.Fr_{H,G}(\alpha) = \Sigma(vg_i^{-1}\alpha) \otimes g_i$ (here, the g_i run over a set of right coset representatives of H in G, and the resulting map is independent of choice of coset representatives)

(ii) $Fr'_{H,G} : (W, V{\downarrow}_H)^H \cong (W{\uparrow}^G, V)^G$
given by $(w \otimes g)Fr'_{H,G}(\beta) = (w\beta)g$. □

Note that this may be interpreted as saying that induction is both left and right adjoint to restriction.

2.1.4 Lemma (Mackey Decomposition)

Suppose H and K are subgroups of G, V is a ΓH-module and W is a ΓK-module.

(i) $V{\uparrow}^G{\downarrow}_K \cong \bigoplus_{HgK} V^g{\downarrow}_{H^g \cap K}{\uparrow}K$. In this expression, g runs over a set of H-K double coset representatives in G, and V^g denotes the H^g-module conjugate to V by g.

(ii) $V{\uparrow}^G \otimes W{\uparrow}^G \cong \bigoplus_{HgK} (V^g{\downarrow}_{H^g \cap K} \otimes W{\downarrow}_{H^g \cap K}){\uparrow}^G$. □

2.1.5 Lemma

If V_1 and V_2 are ΓG-modules and V_1 is projective, then so is $V_1 \otimes V_2$.

Proof

Use 1.4.1 (iii). □

2.1.6 Lemma

(i) $(V,W)^G \cong (\Gamma, \mathrm{Hom}_\Gamma(V,W))^G \cong (\Gamma, V^* \otimes W)^G$

$\cong (\mathrm{Hom}_\Gamma(W,V),\Gamma)^G \cong (V \otimes W^*, \Gamma)^G$

(ii) $(U \otimes V,W)^G \cong (U,V^* \otimes W)^G$. □

Exercises

1. Let G be a p-group and k a field of characteristic p. Let V be a kG-module. Show that G fixes pointwise some nontrivial subspace of V. (Hint: first do the case where G is abelian, and then use the fact that every proper subgroup of G is properly contained in its normalizer). Deduce that there is only one irreducible kG-module, and that the regular representation of the group algebra is indecomposable. What do the decomposition matrix and the Cartan matrix look like?

2. Let G be S_3, the group of all permutations of three objects, and R be the ring of 2-adic integers. Write down the character table for $\hat{R}G$. Find R-forms for each of the irreducible $\hat{R}G$-modules (cf. 1.7.1). Examine their reductions modulo (2). What are the 2-modular irreducible representations (i.e. the irreducible $\bar{R}G$-modules)? Write down the decomposition matrix and Cartan matrix. Deduce that $\bar{R}G$ is isomorphic to the direct sum of $\mathrm{Mat}_2(\bar{R})$ and $\bar{R}H$, where H is a cyclic group of order 2.

2.2 Representation Rings

Since ΓG has the unique decomposition property (see section 2.1), we may define $a(G) = a_\Gamma(G)$ to be the free abelian group on the set of indecomposable ΓG-modules, with multiplication defined on the generators by the tensor product, and then extended bilinearly to $a(G)$ (see 2.1.1). We then let $A(G) = A_\Gamma(G) = a_\Gamma(G) \otimes_{\mathbb{Z}} \mathbb{C}$. $A(G)$ is called the representation ring or Green ring (after J.A. Green) of ΓG. It is a commutative and associative algebra over $\mathbb{C}$, and the identity element is the trivial ΓG-module Γ, which we henceforth write as 1. As we shall see later, the ring $A(G)$ is in general infinite dimensional, and for $\Gamma \in \{R,\bar{R}\}$ it is finite dimensional if and only if the Sylow p-subgroups of G $(p = \mathrm{char}(\bar{R}))$ are cyclic. If H is a subgroup of G, we define $A_0(G,H)$ to be the linear span in A(G) of the elements of the form $X - X' - X''$, where $0 \to X' \to X \to X'' \to 0$ is a short exact sequence which splits on restriction to H (an H-split sequence). Then $A_0(G,H)$ is an ideal of A(G), and $A(G)/A_0(G,1)$ is the Grothendieck ring of ΓG.

We shall be interested in studying the representation theory of G via the study of the structure of a(G) and A(G). We want to look at various subrings and ideals in these rings, and see how they reflect the structure of the group. Each subring or ideal which we define will have an integral version denoted by a small letter a(-), a p-local version $a(-) \otimes_{\mathbb{Z}} \mathbb{Z}(\frac{1}{p}) = \hat{a}(-)$, and a complex version $a(-) \otimes_{\mathbb{Z}} \mathbb{C} = A(-)$.

The first question we may wish to ask is, how are the elements of G reflected in the structure of A(G)? Over a splitting field of characteristic zero, the answer to this question is easy; the columns of the ordinary character table are in one-one correspondence with the conjugacy classes of elements of G. Over a field of characteristic p, the answer is a little more difficult. Brauer discovered that there were certain algebra homomorphisms (see section 2.11), from A(G) to $\mathbb{C}$ which correspond to the conjugacy classes of p'-elements of G. This motivates the following definition.

If A is a subalgebra or ideal of A(G), a species of A is an algebra homomorphism $A \to \mathbb{C}$. If s is a species and $x \in A$, we write (s,x) for the value of s on x.

2.2.1 Lemma

Any set of distinct species of A is linearly independent.

Proof

Suppose $\sum_{i=1}^{r} a_i s_i = 0$ is a linear relation among the species of A with r minimal. Choose $y \in A$ such that $(s_1,y) \neq (s_2,y)$. Then $0 = \sum_{i=1}^{r} a_i(s_i,x.y) = \sum_{i=1}^{r} a_i(s_i,y)(s_i,x)$ and so

$\sum_{i=2}^{r} a_i((s_i,y) - (s_1,y))s_i = 0$. This contradicts the minimality of r. □

If A' is a finite dimensional semisimple subalgebra of A with species $s_1, \ldots, s_r$, lemma 2.2.1 tells us that we may find elements $e_1, \ldots, e_r$ such that $(s_i,e_j) = \delta_{ij}$. Then every species has the same value on e_i^2 as on e_i, and zero on e_ie_j, and so the e_i are primitive orthogonal idempotents. This gives a direct sum decomposition

$$A = \bigoplus_{i=1}^{r} Ae_i \quad .$$

Thus every species of A is a species of some Ae_i and is zero on the Ae_j, $j \neq i$.

If H is a subgroup of G, we get a ring homomorphism $r_{G,H}: A(G) \to A(H)$ by restriction of representations, and a linear map (which is not a ring homomorphism in general) $i_{H,G}: A(H) \to A(G)$ given by induction of representations.

2.2.2 Theorem

If H is a subgroup of G, then

(i) $A(G) = \mathrm{Im}(i_{H,G}) \oplus \mathrm{Ker}(r_{G,H})$ as a direct sum of ideals.

(ii) $A(H) = \mathrm{Im}(r_{G,H}) \oplus \mathrm{Ker}(i_{H,G})$ as a direct sum of vector spaces.

Proof

(i) It follows from 2.1.2 that $\mathrm{Im}(i_{H,G})$ and $\mathrm{Ker}(r_{G,H})$ are ideals. We proceed by induction on $|H|$. If $|H| = 1$, then $\mathrm{codim}(\mathrm{Ker}(r_{G,1})) = \dim(\mathrm{Im}(i_{1,G})) = 1$. Since $i_{1,G}(1) \notin \mathrm{Ker}(r_{G,1})$ the result follows. So suppose $|H| > 1$, and that for any $K < H$, $A(G) = \mathrm{Im}(i_{K,G}) + \mathrm{Ker}(r_{G,K})$. Then $A(G) = \sum_{K<H} \mathrm{Im}(i_{K,G}) + \bigcap_{K<H} \mathrm{Ker}(r_{G,K})$, and so $\mathrm{Im}(r_{G,H}) = r_{G,H}(\sum_{K<H} \mathrm{Im}(i_{K,G})) + \bigcap_{K<H} \mathrm{Ker}_{\mathrm{Im}(r_{G,H})}(r_{H,K})$. Let $1 = a + b$ in this decomposition. Then since $b = 1-a$ is invariant under $N_G(H)$, we have, by the Mackey decomposition theorem (2.1.4), $b{\uparrow^G}{\downarrow_H} = |N_G(H):H|.b$, and so $b \in r_{G,H}(\mathrm{Im}(i_{H,G}))$. Hence $\mathrm{Im}(r_{G,H}) = r_{G,H}(\mathrm{Im}(i_{H,G}))$. Now if $x \in A(G)$, choose $y \in \mathrm{Im}(i_{H,G})$ with $x{\downarrow_H} = y{\downarrow_H}$. Then $x = y + (x-y) \in \mathrm{Im}(i_{H,G}) + \mathrm{Ker}(r_{G,H})$.

Now write $1 = a' + b'$ in this decomposition. If $x \in \mathrm{Im}(i_{H,G}) \cap \mathrm{Ker}(r_{G,H})$, then $x = xa' + xb' = 0$.

(ii) Write $A_1 = \mathrm{Im}(r_{G,H})$ and $A_2 = \mathrm{Ker}(i_{H,G})$. We show by induction on $|K|$, for subgroups $K \le H$, that if M is a ΓK-module, then $M{\uparrow^H} \in A_1 + A_2$. By the Mackey theorem,

$$M{\uparrow^G}{\downarrow_H} = \sum_{KgH} M^g{\downarrow_{K^g \cap H}}{\uparrow^H} \quad .$$

If $K^g \le H$, then $M^g{\downarrow}_{K^g \cap H}{\uparrow}^H \equiv M{\uparrow}^H \mod A_2$.

If $K^g \not\le H$, then $M^g{\downarrow}_{K^g \cap H}{\uparrow}^H \in A_1 + A_2$ by the inductive hypothesis.

Since $M{\uparrow}^G{\downarrow}_H \in A_1$, some positive multiple of $M{\uparrow}^H$ is in $A_1 + A_2$, and hence so is $M{\uparrow}^H$.

Now suppose $x \in A_1 \cap A_2$, with $x = u{\downarrow}_H$. By (i), we may assume that $u \in \mathrm{Im}(i_{H,G})$. Let $e{\uparrow}^{G^2}$ be the idempotent generator of $\mathrm{Im}(i_{H,G})$. Then $u = u.e{\uparrow}^G = (u{\downarrow}_H.e){\uparrow}^G = (x.e){\uparrow}^G$. Write $e = v{\downarrow}_H + w$ with $v{\downarrow}_H \in A_1$ and $w \in A_2$. Then

$$u = (x.v{\downarrow}_H){\uparrow}^G + (x.w){\uparrow}^G = x{\uparrow}^G.v + u.w{\uparrow}^G = 0. \quad \square$$

2.2.3 <u>Corollary</u>

Let $H \le G$ and let V_1 and V_2 be ΓH-modules, and W_1 and W_2 be ΓG-modules.

(i) If $V_1{\uparrow}^G{\downarrow}_H \cong V_2{\uparrow}^G{\downarrow}_H$ then $V_1{\uparrow}^G \cong V_2{\uparrow}^G$.

(ii) If $W_1{\downarrow}_H{\uparrow}^G \cong W_2{\downarrow}_H{\uparrow}^G$ then $W_1{\downarrow}_H \cong W_2{\downarrow}_H$.

<u>Proof</u>

(i) $V_1{\uparrow}^G - V_2{\uparrow}^G \in \mathrm{Im}(i_{H,G}) \cap \mathrm{Ker}(r_{G,H}) = 0$ by 2.2.2(i).

(ii) $W_1{\downarrow}_H - W_2{\downarrow}_H \in \mathrm{Im}(r_{G,H}) \cap \mathrm{Ker}(i_{H,G}) = 0$ by 2.2.2(ii). $\square$

<u>Exercises</u>

1. Let G be cyclic of order p^n, and k a field of characteristic p. Show that there are p^n isomorphism classes of indecomposable kG-modules $V_1, .., V_{p^n}$ with $\dim(V_i) = i$, and $\dim_k \mathrm{Hom}_{kG}(V_i,V_j) = \min(i,j)$. If $x = \Sigma a_i V_i$ and $y = \Sigma b_i V_i$, define $(x,y) = \Sigma a_i b_j \dim_k \mathrm{Hom}_{kG}(V_i,V_j)$. For $x \in (a(G) \otimes \mathbb{R}) \setminus \{0\}$, show that $(x,x) > 0$. Deduce that $a(G) \otimes_{\mathbb{Z}} \mathbb{R}$ is semisimple, and hence that $A(G)$ is semisimple. (See also exercise 3 of 2.18).

2. Now let G be cyclic of order p. Then

$$V_2 \otimes V_i \cong \begin{cases} V_{i+1} \oplus V_{i-1} & \text{if } i \ne p \\ V_p \oplus V_p & \text{if } i = p\,. \end{cases}$$

Find the species of $A(G)$.

3. Let G be S_3, the symmetric group on 3 letters, and let k be a field of characteristic 3. What are the simple kG-modules? What are the indecomposable kG-modules? Draw up a table of tensor

products of indecomposable modules, and find all species of $A(G)$. Display the answer in the form of a 'representation table', with rows indexed by the indecomposable modules and columns indexed by the species. Deduce that $A(G)$ is semisimple. For each subgroup H of G, identify $\mathrm{Im}(i_{H,G})$ and $\mathrm{Ker}(r_{G,H})$. For each species s, find the minimal subgroups H with $\mathrm{Ker}(r_{G,H}) \leq \mathrm{Ker}(s)$. Show that there are numbers $\lambda_i \in \mathbb{C}$ corresponding to the species s_i, with the property that for any representations V and W, we have

$$\dim_k \mathrm{Hom}_{kG}(V,W) = \sum_i (s_i,V)(s_i,W^*)/\lambda_i .$$

For each indecomposable module V_j, find an element $H_j \in A(G)$ with

$$\sum_i (s_i,V_j)(s_i,H_k^*)/\lambda_i = \delta_{jk} .$$

Show that each H_j is of the form $X - X' - X''$, for some short exact sequence $0 \to X' \to X \to X'' \to 0$ of kG-modules.

4. Let k_1 be an algebraic extension of k. Show that the natural map $A_k(G) \to A_{k_1}(G)$ given by $V \mapsto V \otimes_k k_1$ is an injection. (This is a special case of the Noether-Deuring theorem, see for example Curtis and Reiner [38, p.139]. We shall give an unusual proof of this theorem in section 2.18).

5. Let G be a finite group. Define $b(G)$ to be the <u>Burnside Ring</u> of G, namely the free abelian group on the set of transitive permutation representations, with multiplication defined by forming the Cartesian product and decomposing into orbits. Let $B(G) = b(G) \otimes_{\mathbb{Z}} \mathbb{C}$.

(i) For each conjugacy class of subgroup H of G, find a species $s_H\colon B(G) \to \mathbb{C}$ (Hint: look at fixed points). Deduce that $B(G)$ is semisimple and that every species is of the form s_H for some $H \leq G$.

(ii) Show that for every H, $s_H(b(G))$ lies in $\mathbb{Z}$.

(iii) Show that $\prod_{(H)} s_H : b(G) \to \prod_{(H)} \mathbb{Z}$ (the products are taken over conjugacy classes of subgroups H) is an injection whose image is a subgroup of index $\prod_{(H)} |N_G(H):H|$.

(iv) Show that the primitive idempotents in $b(G)$ are in one-one correspondence with the perfect subgroups of G. In particular G is solvable if and only if 1 is the only idempotent in $b(G)$.

2.3 Relative Projectivity and the Trace Map

Let H be a subgroup of G. Let $\{g_i : i \in I\}$ denote a set of right coset representatives of H in G. Suppose U and V are ΓG-modules. Then we define the trace map

$$\mathrm{Tr}_{H,G}: (U,V)^H \to (U,V)^G \quad \text{via}$$

$$u.\mathrm{Tr}_{H,G}(\varphi) = \sum_{i \in I} ((ug_i^{-1})\varphi)g_i \,.$$

It is clear that the map $\mathrm{Tr}_{H,G}$ is independent of the choice of coset representatives. If we consider $(U,V)^H$ and $(U,V)^G$ as $\mathrm{End}_{\Gamma G}(U)$ - $\mathrm{End}_{\Gamma G}(V)$ bimodules, then $\mathrm{Tr}_{H,G}$ is a bimodule homomorphism (see 2.3.1(i) & (ii)). We write $(U,V)^G_H$ for the image of $\mathrm{Tr}_{H,G}$ and $(U,V)^{H,G}$ for the cokernel. If $\mathcal{H}$ is a collection of subgroups, then we write $(U,V)^G_{\mathcal{H}}$ for the sum of the images of $\mathrm{Tr}_{H,G}$, $H \in \mathcal{H}$, and $(U,V)^{\mathcal{H},G}$ for $(U,V)^G/(U,V)^G_{\mathcal{H}}$. We also write V^G_H, $V^{H,G}$, $V^G_{\mathcal{H}}$ and $V^{\mathcal{H},G}$ for $(\Gamma,V)^G_H$, $(\Gamma,V)^{H,G}$, $(\Gamma,V)^G_{\mathcal{H}}$ and $(\Gamma,V)^{\mathcal{H},G}$ respectively.

2.3.1 Lemma

(i) If $\alpha \in (U,V)^H$ and $\beta \in (V,W)^G$ then $\mathrm{Tr}_{H,G}(\alpha)\beta = \mathrm{Tr}_{H,G}(\alpha\beta)$.

(ii) If $\alpha \in (U,V)^G$ and $\beta \in (V,W)^H$ then $\alpha\mathrm{Tr}_{H,G}(\beta) = \mathrm{Tr}_{H,G}(\alpha\beta)$.

(iii) In particular $(U,U)^G_H$ is an ideal in $\mathrm{End}_{\Gamma G}(U)$.

(iv) If U and W are ΓG-modules and V is a ΓH-module, then for $\alpha \in (U,V)^H$ and $\beta \in (V,W)^H$,

$$\mathrm{Fr}_{H,G}(\alpha)\mathrm{Fr}'_{H,G}(\beta) = \mathrm{Tr}_{H,G}(\alpha\beta).$$

(v) If $L \le H \le G$, then $\mathrm{Tr}_{H,G}(\mathrm{Tr}_{L,H}(\alpha)) = \mathrm{Tr}_{L,G}(\alpha)$.

(vi) If H and K are two subgroups of G, then for $\alpha \in \mathrm{Hom}_{\Gamma H}(U,V)$, $\mathrm{Tr}_{H,G}(\alpha) = \sum_{HgK} \mathrm{Tr}_{H^g \cap K,K}(\alpha g)$.

(vii) If $\alpha \in \mathrm{End}_{\Gamma H}(U)$ and $\beta \in \mathrm{End}_{\Gamma K}(U)$, then

$$\mathrm{Tr}_{H,G}(\alpha)\mathrm{Tr}_{K,G}(\beta) = \sum_{HgK} \mathrm{Tr}_{H^g \cap K,G}(\alpha g \beta) \,.$$

Proof

(i) - (v) are clear from the definition.

(vi) For each double coset HgK, let $\lambda(g)$ be a set of right coset representatives of $H^g \cap K$ in K. Then $\bigcup_{HgK} \{gk : k \in \lambda(g)\}$ is a set of right coset representatives of H in G.

(vii) $\mathrm{Tr}_{H,G}(\alpha)\mathrm{Tr}_{K,G}(\beta) = \mathrm{Tr}_{K,G}(\mathrm{Tr}_{H,G}(\alpha)\beta)$ by (ii)

$= \mathrm{Tr}_{K,G}(\sum_{HgK} \mathrm{Tr}_{H^g\cap K,K}(\alpha g)\beta)$ by (vi)

$= \mathrm{Tr}_{K,G}(\sum_{HgK} \mathrm{Tr}_{H^g\cap K,K}(\alpha g\beta))$ by (i)

$= \sum_{HgK} \mathrm{Tr}_{H^g\cap K,G}(\alpha g\beta)$ by (v). □

A module V is said to be <u>H-projective</u> or <u>projective relative to</u> <u>H</u> if whenever we have a map $\lambda: V \to X$, and a surjection $\mu: W \to X$ such that as ΓH-modules there is a map $\nu: V \to W$ and $\lambda=\nu\mu$, then there is also a ΓG-module homomorphism ν' with $\lambda = \nu'\mu$.

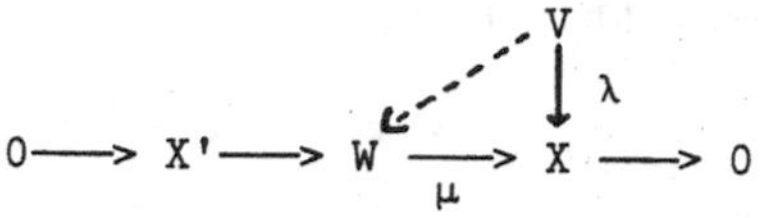

2.3.2 <u>Proposition</u> (D. G. Higman's lemma)

Let V be a ΓG-module and H a subgroup of G. The following are equivalent.

(i) V is H-projective.

(ii) V is a summand of $V{\downarrow_H}{\uparrow^G}$.

(iii) V is a summand of $U{\uparrow^G}$ for some ΓH-module U.

(iv) $1_V \in \mathrm{Im}(\mathrm{Tr}_{H,G}: \mathrm{End}_{\Gamma H}(V) \to \mathrm{End}_{\Gamma G}(V))$.

<u>Proof</u>

(i) ⇒ (ii): The defining condition implies that the natural surjection $\mathrm{Fr}'_{H,G}(1_{V\downarrow_H}): V{\downarrow_H}{\uparrow^G} \to V$ splits.

(ii) ⇒ (iii) is clear.

(iii) ⇒ (iv): Let $\rho = \mathrm{Fr}^{-1}_{H,G}(1_{U\uparrow^G}).\mathrm{Fr}'^{-1}_{H,G}(1_{U\uparrow^G}) \in \mathrm{End}_{\Gamma H}(U{\uparrow^G}{\downarrow_H})$. Then by 2.3.1(iv), $\mathrm{Tr}_{H,G}(\rho) = 1_{U\uparrow^G}$. Thus if

$$\theta : V{\downarrow_H} \hookrightarrow U{\uparrow^G}{\downarrow_H} \xrightarrow[\rho]{} U{\uparrow^G}{\downarrow_H} \twoheadrightarrow V{\downarrow_H}$$

then $\mathrm{Tr}_{H,G}(\theta) = 1_V$ by 2.3.1(i) & (ii).

(iv) ⇒ (i): Let $\lambda: V \to X$ and $\mu: W \to X$ and $\nu: V \to W$ as in the definition of H-projective. Let $\theta \in \mathrm{End}_{\Gamma H}(V)$ with $\mathrm{Tr}_{H,G}(\theta) = 1_V$, and let $\nu' = \mathrm{Tr}_{H,G}(\theta\nu)$. Then

$$\nu'\mu = \mathrm{Tr}_{H,G}(\theta\nu)\mu = \mathrm{Tr}_{H,G}(\theta\nu\mu) = \mathrm{Tr}_{H,G}(\theta\lambda) = \mathrm{Tr}_{H,G}(\theta)\lambda = \lambda . \quad \square$$

2.3.3 <u>Corollary</u>

Suppose $\Gamma \in \{\hat{R},R,\bar{R}\}$. If $P \in \mathrm{Syl}_p(G)$, then every ΓG-module V is P-projective.

Proof

For each of $\Gamma = \hat{R}$, R or $\bar{R}$, $\frac{1}{|G:P|}$ exists in Γ, and so $1_V = \mathrm{Tr}_{P,G}(\frac{1}{|G:P|}.1_V)$. □

2.3.4 Corollary

Let $\alpha \in (U,V)^G$. Then the following are equivalent.

(i) $\alpha \in (U,V)^G_H$

(ii) There are an H-projective module W and maps $\beta: U \to W$ and $\gamma: W \to V$ such that $\alpha = \beta\gamma$.

Proof

(i) ⇒ (ii): Take $W = V{\downarrow_H}{\uparrow^G}$, and let $\alpha = \mathrm{Tr}_{H,G}(\alpha')$. Let $\beta = \mathrm{Fr}_{H,G}(\alpha')$ and $\gamma = \mathrm{Fr}'_{H,G}(1_{V{\downarrow_H}})$. Then

$$\beta\gamma = \mathrm{Fr}_{H,G}(\alpha')\mathrm{Fr}'_{H,G}(1_{V{\downarrow_H}}) = \mathrm{Tr}_{H,G}(\alpha').$$

(ii) ⇒ (i) : Suppose W is a summand of $X{\uparrow^G}$. Let $\beta': U \xrightarrow{\beta} W \hookrightarrow X{\uparrow^G}$ and $\gamma': X{\uparrow^G} \twoheadrightarrow W \xrightarrow{\gamma} V$. Then $\mathrm{Tr}_{H,G}(\mathrm{Fr}^{-1}_{H,G}(\beta')\mathrm{Fr}'^{-1}_{H,G}(\gamma')) = \beta'\gamma' = \alpha$ by 2.3.1(iv). □

2.3.5 Corollary

If V, W are ΓG-modules and V is H-projective then so is $V \otimes W$.

Proof

Suppose V is a summand of $X{\uparrow^G}$. Then $V \otimes W$ is a summand of $X{\uparrow^G} \otimes W = (X \otimes W{\downarrow_H}){\uparrow^G}$. □

Remark

Suppose W is a submodule of V, and W and V/W are projective. Then V is also projective, since it is isomorphic to a direct sum of W and V/W. However, the same is not true if the word projective is replaced by H-projective. For example, let G be the cyclic group of order four, and let H be the subgroup of order two. Then there are four isomorphism classes V_1, V_2, V_3 and V_4 of indecomposable kG-modules, for k the field of two elements. Their dimensions are 1, 2, 3 and 4 respectively. V_2 and V_4 are H-projective, while V_1 and V_3 are not. However, there is a short exact sequence

$$0 \longrightarrow V_2 \longrightarrow V_1 \oplus V_3 \longrightarrow V_2 \longrightarrow 0.$$

Now let G act as permutations on a finite set S, and denote by S^G the set of fixed points of G on S. Let $\mathrm{Fix}_G(S) = \{H \le G: S^H \neq \emptyset\}$, and denote by ΓS the ΓG-permutation module corresponding to S. An exact sequence $0 \to X' \to X \to X'' \to 0$ is said to be <u>S-split</u> if the sequence

$$0 \to X' \otimes \Gamma S \to X \otimes \Gamma S \to X'' \otimes \Gamma S \to 0 \quad \text{splits.}$$

2.3.6 Lemma

A sequence $0 \to X' \to X \overset{\mu}{\to} X'' \to 0$ is S-split if and only if it splits on restriction to every $H \in \mathrm{Fix}_G(S)$.

Proof

If we write $S = \dot{\bigcup} S_i$ as a sum of orbits of G, then the sequence is S-split if and only if it is S_i-split for each i, so we may restrict our attention to the transitive case. Let H be the stabilizer of a point z_o in S. If the sequence splits on restriction to H, then it is clearly S-split since $X \otimes \Gamma S = X{\downarrow_H}{\uparrow^G}$. Conversely suppose $f: X'' \otimes \Gamma S \to X \otimes \Gamma S$ is an S-splitting. For $x \in X''$, write $(x \otimes z_o)f = \sum_{z_i \in S} y_i \otimes z_i$, and let $xf_o = y_o$. Then $f_o: X'' \to X$ is an H-splitting, since $f_o\mu = 1$ and $(xh)f_o = (xh \otimes z_o)f = ((x \otimes z_o)h)f = y_oh = (xf_o)h$. □

We say that a module V is <u>S-projective</u> or <u>projective relative to S</u> if whenever we have a homomorphism $W \underset{\mu}{\to} V \to 0$ which is S-split, then μ is split. Thus if S is transitive with stabilizer H, then V is S-projective if and only if it is H-projective. In general, V is S-projective if and only if it is a sum of H-projective modules for $H \in \mathrm{Fix}_G(S)$, which by 2.3.2 happens if and only if it is a direct summand of $V \otimes \Gamma S$.

2.3.7 Lemma

If V and W are ΓG-modules and V is S-projective then so is $V \otimes W$.

Proof

This follows from 2.3.5. □

We denote by $A(G,H)$ the ideal of **A(G) spanned by the H-projective** modules (see 2.3.5), $A(G,S)$ the ideal spanned by the S-projective modules (see 2.3.7), and $A_o(G,H)$ (resp. $A_o(G,S)$) the ideal spanned by the elements of the form $X - X' - X''$ where $0 \to X' \to X \to X'' \to 0$ is an H-split (resp. S-split) sequence. If $\mathcal{H}$ is a collection of subgroups, we write $A(G,\mathcal{H})$ for the ideal spanned by the $A(G,H)$ for $H \in \mathcal{H}$, and $A_o(G,\mathcal{H})$ for the intersection of the $A_o(G,H)$ for $H \in \mathcal{H}$. Then clearly $A(G,S) = A(G,\mathrm{Fix}_G(S))$ and $A_o(G,S) = A_o(G,\mathrm{Fix}_G(S))$. We also write $A'(G,H)$ for $A(G,\mathcal{H})$ where $\mathcal{H}$ is the set of proper subgroups of H.

2.3.8 Lemma

Suppose H is a subgroup of G and G acts as permutations on S. We let H act on S by restriction of the action of G. Then

(i) $i_{H,G}(A(H,S)) \subseteq A(G,S)$

(ii) $i_{H,G}(A_o(H,S)) \subseteq A_o(G,S)$

(iii) $r_{G,H}(A(G,S)) \subseteq A(H,S)$

(iv) $r_{G,H}(A_o(G,S)) \subseteq A_o(H,S)$

The Proof is an easy exercise. □

We shall see in section 2.15 that in fact for any permutation representation S of G,

$$A(G) = A(G,S) \oplus A_o(G,S).$$

Exercise

Show that V is a projective kG-module if and only if $V^* \otimes V$ is projective. Deduce that $V \otimes W$ is projective if and only if $V^* \otimes W$ is projective. [Hint: reduce to a Sylow p-subgroup and use 2.3.2].

2.4 The Inner Products on A(G)

We define two different bilinear forms (,) and < , > on A(G) as follows.

If V and W are ΓG-modules, we let

$$(V,W) = \text{rank}_\Gamma \text{Hom}_{\Gamma G}(V,W) = \text{rank}_\Gamma \text{Hom}_{\Gamma G}(1, V^* \otimes W) \text{ by 2.1.1(iv).}$$

We then extend (,) bilinearly to give a (not necessarily symmetric) bilinear form on the whole of A(G).
Note that $(xy,z) = (x,y^*z)$, by 2.1.6 (ii).

For $\Gamma = \hat{R}$ or $\Gamma = R$, the bilinear form (,) is symmetric. However, for a field k of characteristic p, (,) is not necessarily symmetric. We thus introduce another inner product

$$\langle V,W \rangle = \dim_k(V,W)^G_1 \quad \text{(see 2.3)}$$
$$= \dim_k(1, V^* \otimes W)^G_1 \quad \text{by 2.1.1(iv)}$$

and extend bilinearly to the whole of $A_k(G)$.

Note that $\langle xy,z \rangle = \langle x,y^*z \rangle$. The relationship between the two inner products (,) and < , > is given in 2.4.3.

Let $P_1 = (P_1)_{kG}$ be the projective cover of the trivial kG-module 1, and let

$$u = u_{kG} = P_1 - \Omega^{-1}(1)$$
$$v = v_{kG} = P_1 - \Omega(1)$$

as elements of $A(G)$ (note that $u^* = v$).

2.4.1 Lemma

The following expressions are equal.

(i) $< V,W >$.

(ii) The multiplicity of $(P_1)_{kG}$ as a direct summand of $\mathrm{Hom}_k(V,W) = V^* \otimes W$.

(iii) $(u,\mathrm{Hom}_k(V,W))$.

(iv) The rank of $\sum_{g \in G} g$ in the matrix representation of kG on $\mathrm{Hom}_k(V,W)$.

In particular, $< \ , \ >$ is symmetric.

Proof

Since each of these expressions is unaffected by replacing V by 1 and W by $\mathrm{Hom}_k(V,W)$, we may assume that $V = 1$. Also, since each expression is additive in W, we may suppose that W is indecomposable. We shall now show that each of these expressions is 1 when $W \cong P_1$ and zero otherwise.

(i) If $< 1,W > \neq 0$, then there is an element $\alpha \in W$ with $\sum_{g \in G} \alpha g \neq 0$. Let $\lambda: P \to W$ be the projective cover of W, and write $P = nP_1 \oplus P'$, where P' has no direct summands isomorphic to P_1. Choose an element $\alpha_1 + \alpha_2 \in nP_1 \oplus P'$ with $(\alpha_1 + \alpha_2)\lambda = \alpha$. Then $\sum_{g \in G} (\alpha_1 + \alpha_2)g \neq 0$. But $\sum_{g \in G} \alpha_2 g = 0$, since P' has no invariant elements, and so $(\sum_{g \in G} \alpha_1 g)\lambda \neq 0$. Thus the submodule spanned by α_1 is a copy of P_1 whose socle $\sum_{g \in G} \alpha_1 g$ is not killed by λ. Since P_1 is injective, this means that P_1 is a direct summand of W, and since W is indecomposable, $P_1 \cong W$. Clearly $< 1,P_1 > = 1$.

(ii) This is clear.

(iii) A homomorphism from P_1 to W factors through $\Omega^{-1}(1)$ unless $W \cong P_1$, since P_1 is injective. Thus if $W \not\cong P_1$, $(u,W) = 0$. On the other hand, if $W \cong P_1$, then any homomorphism $P_1 \to W$ is equivalent modulo a multiple of this isomorphism to a homomorphism factoring through $\Omega^{-1}(1)$, and so $(u,W) = 1$.

(iv) This is clearly the same as (i).

The symmetry of $< \ , \ >$ follows since $P_1 \cong P_1^*$. □

2.4.2 <u>Proposition</u>

Let V be an indecomposable kG-module with projective cover P_V and injective hull I_V. Then we have the following :

(i) $\Omega^{-1}(1) \otimes \Omega(V) \cong V \oplus$ projectives.

(ii) $u.(P_V - \Omega(V)) = V = v.(I_V - \Omega^{-1}(V))$
and in particular $u.v = 1$.

(iii) $u.V = I_V - \Omega^{-1}(V)$
$v.V = P_V - \Omega(V)$.

<u>Proof</u>

We have short exact sequences $0 \to 1 \to P_1 \to \Omega^{-1}(1) \to 0$ and $0 \to \Omega(V) \to P_V \to V \to 0$. Tensor the first of these with V, and the second with $\Omega^{-1}(1)$. Then applying Schanuel's lemma (1.4.2), we get

$$(*) \qquad \Omega^{-1}(1) \otimes \Omega(V) \oplus P_1 \otimes V \cong \Omega^{-1}(1) \otimes P_V \oplus V,$$

which proves (i).

Thus as elements of $A(G)$, we get

$$u.(P_V - \Omega(V)) = P_1.P_V - (P_1.P_V - P_1.V) - \Omega^{-1}(1).P_V + \Omega^{-1}(1).\,\Omega(V)$$

(note that $P_1.\Omega(V) = P_1.P_V - P_1.V$ since $P_1.V$ is projective by 2.1.5)

$$= V, \text{ by } (*) \text{ above.}$$

This statement and its dual prove (ii), and (iii) follows immediately. □

2.4.3 <u>Corollary</u>

Let V and W be kG-modules. Then

(i) $(V,W) = \langle v.V,W\rangle = \langle v,\mathrm{Hom}_k(V,W)\rangle$
$= \langle V,u.W\rangle = \langle \mathrm{Hom}_k(W,V),u\rangle$.

(ii) $\langle V,W\rangle = (u.V,W) = (u,\mathrm{Hom}_k(V,W))$
$= (V,v.W) = (\mathrm{Hom}_k(W,V),v)$.

(iii) $(V,W) = (W,v^2.V)$.

<u>Proof</u>

$\langle V,W\rangle = (u,\mathrm{Hom}_k(V,W))$ by 2.4.1(iii)
$= (u.V,W)$ by 2.1.6(i).

The rest follow similarly from 2.1.6 and the fact that $u.v = 1$ (2.4.2(ii)). □

2.4.4 <u>Corollary</u>

Let V and W be kG-modules.

If V is indecomposable and W is irreducible, then

$$< V,W > = \begin{cases} \dim_k \mathrm{End}_{kG}(W) & \text{if } V \cong P_W \\ 0 & \text{otherwise.} \end{cases}$$

Proof

$< V,W > = (V,v.W) = (V,P_W - \Omega(W))$ by 2.4.3 and 2.4.2(iii). Since V is indecomposable and P_W is projective, any homomorphic image of V in P_W lies in $\Omega(W)$ unless $V \cong P_W$. □

2.4.5 Proposition

If $H \leq G$, then

(i) $u_{kG}{\downarrow}_H = u_{kH}$

(ii) $v_{kG}{\downarrow}_H = v_{kH}$.

Proof

We have short exact sequences $0 \to \Omega(1)_{kH} \to (P_1)_{kH} \to 1_{kH} \to 0$ and $0 \to \Omega(1)_{kG}{\downarrow}_H \to (P_1)_{kG}{\downarrow}_H \to 1_{kH} \to 0$. Thus (ii) follows from Schanuel's lemma (1.4.2) and (i) is proved dually. □

2.4.6 Corollary

If V is a kH-module and W is a kG-module, then

$$< V,W{\downarrow}_H > = < V{\uparrow}^G,W >$$

(cf. 2.1.3, Frobenius Reciprocity)

Proof

$$\begin{aligned} < V,W{\downarrow}_H > &= (u_{kH},\mathrm{Hom}_k(V,W{\downarrow}_H)) && \text{by 2.4.1} \\ &= (u_{kG}{\downarrow}_H,\mathrm{Hom}_k(V,W{\downarrow}_H)) && \text{by 2.4.5(i)} \\ &= (u_{kG},(\mathrm{Hom}_k(V,W{\downarrow}_H)){\uparrow}^G) && \text{by 2.1.3(i)} \\ &= (u_{kG},\mathrm{Hom}_k(V{\uparrow}^G,W)) && \text{by 2.1.2(vi)} \\ &= < V{\uparrow}^G,W > && \text{by 2.4.1.} \end{aligned}$$ □

2.5 Vertices and Sources

Let V be an indecomposable ΓG-module. Then D is a vertex of V if V is D-projective, but not D'-projective for any proper subgroup D' of D. A source of V is an indecomposable ΓD-module W, where D is a vertex of V, such that V is a direct summand of $W{\uparrow}^G$ (c.f. 2.3.2).

2.5.1 Proposition

Suppose $\Gamma \in \{R,\bar{R}\}$.

Let V be an indecomposable ΓG-module.

(i) The vertices of V are conjugate p-subgroups of G.

(ii) Let W_1 and W_2 be two ΓD-modules which are sources of V. Then there is an element $g \in N_G(D)$ with $W_1 \cong W_2{}^g$.

Proof

(i) Let D_1 and D_2 be vertices of V. Write

$$1_V = 1_V{}^2 = Tr_{D_1,G}(\alpha)Tr_{D_2,G}(\beta) \quad \text{by 2.3.2}$$

$$= \sum_{D_1gD_2} Tr_{D_1{}^g \cap D_2,G}(\alpha g\beta) \quad \text{by 2.3.1 (vii)}$$

$$\in \sum_{D_1gD_2} (V,V)^G_{D_1{}^g \cap D_2} .$$

Thus by Rosenberg's lemma (1.5.5) and minimality of D_1 and D_2, for some $g \in G$ we have $D_1{}^g = D_2$. By 2.3.3, the vertices are p-groups.

(ii) Let W be an indecomposable summand of $V{\downarrow}_D$ which is a source of V (cf. 2.3.2). Then W is also a summand of $W_1{\uparrow}^G{\downarrow}_D = \bigoplus_{DgD} W_1{}^g{\downarrow}_{D^g \cap D}{\uparrow}^D$. Thus for some $g \in N_G(D)$, $W \cong W_1{}^g$. □

If s is a species of $A_\Gamma(G)$, we define a vertex of s to be a vertex of minimal size over indecomposable modules V for which $(s,V) \neq 0$.

2.5.2 Proposition

Suppose $\Gamma \in \{R,\bar{R}\}$.

(i) If W is an indecomposable module with $(s,W) \neq 0$ then every vertex of s is contained in a vertex of W.

(ii) The vertices of s are conjugate p-subgroups of G.

Proof

(i) Suppose D is a vertex of s, and of V with $(s,V) \neq 0$. If $(s,W) \neq 0$, then $(s,V \otimes W) \neq 0$, and so $(s,X) \neq 0$ for some indecomposable direct summand X of $V \otimes W$. But every vertex of X is contained in both a vertex of V and a vertex of W, by 2.3.5. By minimality, D is a vertex of X and is contained in a vertex of W.

(ii) follows immediately from (i). □

Exercises

An algebra has finite representation type if there are only finitely many isomorphism classes of indecomposable modules; otherwise it has *infinite representation type*.

1. Show that the group algebra kP of a group P isomorphic to the direct product of two copies of the cyclic group of order p has infinite representation type (hint: for k infinite, construct an infinite family of non-isomorphic indecomposable two-dimensional representations; then pass down to k finite).
2. (Harder) Show that the group algebra described in exercise 1 has indecomposable representations of arbitrarily large dimension (hint: form an amalgamated sum of copies of $kP/J^2(kP)$, or look at $\Omega^n(k)$).
3. Use the theory of vertices and sources to show that the group algebra kG of a general finite group G has finite representation type if and only if the Sylow p-subgroups of G are cyclic (hint: if a p-group is non-cyclic then it has a quotient isomorphic to the group P of exercise 1). See also 2.12.9.

Remark

The algebras of infinite representation type split further into *tame* and *wild*. Roughly speaking, tame representation type means that the representations are classifiable, whilst a classification of the representations of an algebra of wild representation type would imply a classification of pairs of (non-commuting) matrices up to conjugacy. For a more precise definition, see Ringel's article 'Tame algebras' in 'Representation Theory I', Springer Lecture Notes in Mathematics no. 831, p. 155. For modular group algebras of infinite representation type, it turns out that if $\mathrm{char}(k) \neq 2$, they are all of wild representation type. For $\mathrm{char}(k) = 2$, tame representation type occurs exactly when the Sylow 2-subgroups of G are dihedral, semidihedral, quaternion or generalized quaternion [97].

2.6 Trivial Source Modules

A module V is a *trivial source module* if each indecomposable direct summand has the trivial module as a source.

2.6.1 Lemma

An indecomposable module has trivial source if and only if it is a direct summand of a permutation module.

Proof

If V is a summand of $1_H\uparrow^G$, D is a vertex of V, and the ΓD-module W is a source, then W is a summand of $1_H\uparrow^G\downarrow_D = \bigoplus_{HgD} 1_{H^g\cap D}\uparrow^D$. Since D is a vertex, $W \cong 1_D$. □

We denote by $A(G,Triv)$ the subring of $A(G)$ spanned by the trivial source modules. It turns out that this subring controls many properties of species. We shall investigate this in section 2.14. In this section we shall study the endomorphism ring of a trivial source module, as an example of the theory set up in section 1.7.

2.6.2 Proposition (Scott)

Let V_1 and V_2 be the RG-permutation modules on the cosets of H_1 and H_2. Then the natural map from $Hom_{RG}(V_1,V_2)$ to $Hom_{\overline{R}G}(\overline{V}_1,\overline{V}_2)$ given by reduction modulo (π) is a surjection.

Proof

By the Mackey decomposition theorem, $Hom_{RG}(V_1,V_2)$ and $Hom_{\overline{R}G}(\overline{V}_1,\overline{V}_2)$ have the same rank, namely the number of double cosets H_1gH_2. □

2.6.3 Corollary

(i) Any trivial source $\overline{R}G$-module lifts (uniquely) to a trivial source RG-module.

(ii) If V_1 and V_2 are trivial source RG-modules, then the natural map $Hom_{RG}(V_1,V_2) \to Hom_{\overline{R}G}(\overline{V}_1,\overline{V}_2)$ given by reduction modulo (π) is a surjection.

Proof

Suppose U is a direct summand of an $\overline{R}G$-permutation module $\overline{V}$. By 2.6.2, the natural map $End_{RG}(V) \to End_{\overline{R}G}(V)$ is surjective. Thus by the idempotent refinement theorem (1.7.2), the idempotent corresponding to U lifts to an idempotent in $End_{RG}(V)$, and U is thus the reduction modulo (π) of the corresponding direct summand of V.

It now follows from 2.6.2 that homomorphisms between trivial source $\overline{R}G$-modules lift to homomorphisms between their lifts.

It remains to prove uniqueness of the lift. Suppose W_1 and W_2 are two trivial source lifts of U. Then the identity automorphism of U lifts to maps $W_1 \rightleftarrows W_2$ whose composite either way reduces mod (π) to the identity map. Thus by 1.1.3 the composites are automorphisms, and the maps are hence isomorphisms. □

Remark

2.6.3 may be interpreted as saying that the natural map $A_R(G,Triv) \to A_{\overline{R}}(G,Triv)$ is an isomorphism.

For the remainder of 2.6, we assume that $(\hat{R},R,\overline{R})$ is a splitting p-modular system for RG.

Let G act as permutations on a set S, and let $\hat{R}S$, RS and $\bar{R}S$ be the corresponding $\hat{R}G$, RG and $\bar{R}G$ permutation modules on S. Let $\hat{E} = \hat{E}(S)$, $E = E(S)$ and $\bar{E} = \bar{E}(S)$ denote the endomorphism rings $E_{\Gamma G}(S)$ for $\Gamma = \hat{R}$, R and $\bar{R}$ respectively.

Now 1.2.3 tells us that $\hat{E}(S)$ is semisimple, so that by 2.6.2, $\hat{E}$, E and $\bar{E}$ satisfy the conditions required in section 1.7. Also, by 1.2.3, since $(\hat{R},R,\bar{R})$ is a splitting system for RG, it is also a splitting system for E, and so we have as $\hat{E}$-modules,
$\hat{R}S = \bigoplus_e \dim(V_e).X_e$ where e runs over a set of primitive central idempotents in $\hat{E}$, V_e is the corresponding irreducible $\hat{R}G$-module, and X_e is the corresponding irreducible $\hat{E}$-module. As $\hat{R}G$-modules, we have $\hat{R}S.e = \dim(X_e).V_e$. Indeed, by 1.2.3, as modules for $\hat{E} \otimes_{\hat{R}} \hat{R}G$,

(1) $$\hat{R}S = \bigoplus_e (X_e \otimes V_e).$$

Now let G act on $S \times S$ via the diagonal action $(x,y)g = (xg,yg)$, and write $S \times S = \dot{\bigcup} S_i^2$ as G-orbits. Let $k_i = |S_i^2|$, and let $A^{(i)}$ denote the <u>suborbit map</u> on $\mathbb{Z}S$

$$A^{(i)}: x \to \sum_{(x,y)\in S_i^2} y.$$

Then the $A^{(i)}$ form a $\mathbb{Z}$-basis for $\mathrm{End}_{\mathbb{Z}G}(\mathbb{Z}S)$, and a Γ-basis for $E_{\Gamma G}(S)$ for $\Gamma = \hat{R}$, R or $\bar{R}$. We define a pairing of the suborbits $i \leftrightarrow i'$ via

$$(x,y) \in S_i^2 \Leftrightarrow (y,x) \in S_{i'}^2 .$$

Then it is clear that

(2) $$\mathrm{Tr}_{\Gamma S}(A^{(i')}A^{(j)}) = k_i\delta_{ij} .$$

The following theorem gives the idempotent e in terms of the $A^{(i)}$, an expression for $\dim(V_e)$ and an orthogonality relation.

2.6.4 <u>Theorem</u>

(i) $$e = \dim(X_e).\frac{\sum_i \mathrm{Tr}_{X_e}(A^{(i')})A^{(i)}/k_i}{\sum_i \mathrm{Tr}_{X_e}(A^{(i')})\mathrm{Tr}_{X_e}(A^{(i)})/k_i} .$$

(ii) $$\dim(V_e) = \frac{\dim(X_e)}{\sum_i \mathrm{Tr}_{X_e}(A^{(i')})\mathrm{Tr}_{X_e}(A^{(i)})/k_i} .$$

(iii) If $e \neq e'$ then

$$\sum_i \mathrm{Tr}_{X_{e'}}(A^{(i')})\mathrm{Tr}_{X_e}(A^{(i)})/k_i = 0.$$

Proof

Let $e = \Sigma\, e_i A^{(i)}$. Then by (2),

$$Tr_{\hat{RS}}(A^{(i')}e) = k_i e_i .$$

On the other hand, (1) gives

$$Tr_{\hat{RS}}(A^{(i')}e) = \dim(V_e) Tr_{X_e}(A^{(i')}) .$$

Hence

$$(3) \qquad e_i = \dim(V_e) Tr_{X_e}(A^{(i')})/k_i .$$

But $Tr_{X_e}(e) = \dim(X_e)$, and so

$$\dim(V_e).\sum_i Tr_{X_e}(A^{(i')}) Tr_{X_e}(A^{(i)})/k_i = \dim(X_e).$$

This proves (ii), and substituting back in (3) gives (i). The relation (iii) follows since $Tr_{X_e}(e') = 0$. □

A central component of $\overline{R}S$ is a direct summand of the form $\overline{R}S.e$ where e is a central idempotent in $\overline{E}$. A component of $\overline{R}S$ is an indecomposable direct summand, and corresponds to a primitive idempotent in $\overline{E}$. The central homomorphism ω_e: $Z(E) \to R$ determined by X_e is clearly just

$$\omega_e : \sum_i a_i A^{(i)} \mapsto \sum_i a_i Tr_{X_e}(A^{(i)})/\dim(X_e).$$

Thus two summands of $\hat{R}S$ lie in the same central component of $\overline{R}S$ if and only if the values of the corresponding ω_e are congruent modulo (π) on $Z(E)$. To calculate the components, however, we need information about the decomposition numbers of E.

Example

A_5 acting on the vertices of a dodecahedron

Suborbit maps:

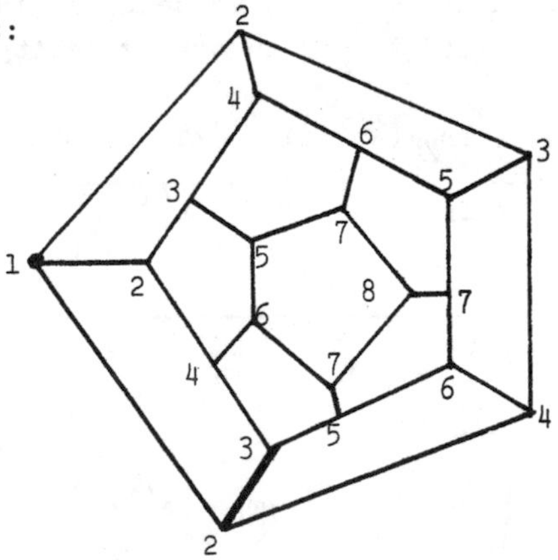

Matrices for the adjoint representation of E

$$A^{(1)}=\begin{pmatrix}1&&&&&&&\\&1&&&&&&\\&&1&&&&&\\&&&1&&&&\\&&&&1&&&\\&&&&&1&&\\&&&&&&1&\\&&&&&&&1\end{pmatrix}\quad A^{(2)}=\begin{pmatrix}&3&&&&&&\\1&&1&1&&&&\\&1&&1&1&&&\\&1&1&&&1&&\\&&1&&&1&1&\\&&&1&1&&1&\\&&&&1&1&&1\\&&&&&&3&\end{pmatrix}$$

$$A^{(3)}=\begin{pmatrix}&&&3&&&&\\&1&1&&1&&&\\1&&&&&2&&\\&1&1&&&&1&\\&&&2&&&&1\\&1&&&1&&1&\\&&1&&1&&1&\\&&&&&3&&\end{pmatrix}\quad A^{(4)}=\begin{pmatrix}&&3&&&&&\\&1&&1&&1&&\\&1&&1&&&1&\\1&&&&2&&&\\&1&&&&1&1&\\&&2&&&&&1\\&&&1&&1&1&\\&&&&3&&&\end{pmatrix}$$

$$A^{(5)}=\begin{pmatrix}&&&&3&&&\\&&&1&&1&1&\\&&2&&&&&1\\&1&&&&1&1&\\1&&&&2&&&\\&1&&1&&&1&\\&1&&1&&1&&\\&&3&&&&&\end{pmatrix}\quad A^{(6)}=\begin{pmatrix}&&&&&3&&\\&&1&&1&&1&\\&1&&&1&&1&\\&&&2&&&&1\\&1&1&&&&1&\\1&&&&&2&&\\&1&1&&1&&&\\&&&3&&&&\end{pmatrix}$$

$$A^{(7)}=\begin{pmatrix}&&&&&&3&\\&&&&1&1&&1\\&&&1&1&&1&\\&&1&&&1&1&\\&1&1&&&1&&\\&1&&1&1&&&\\1&&1&1&&&&\\&3&&&&&&\end{pmatrix}\quad A^{(8)}=\begin{pmatrix}&&&&&&&1\\&&&&&&1&\\&&&&&1&&\\&&&&1&&&\\&&&1&&&&\\&&1&&&&&\\&1&&&&&&\\1&&&&&&&\end{pmatrix}$$

$\hat{E}$ has four one-dimensional representations and a two-dimensional one. The following table gives the matrices for these representations, and the associated characters of A_5.

	Matrix on V_e								Character of X_e				
	$A^{(1)}$	$A^{(2)}$	$A^{(3)}$	$A^{(4)}$	$A^{(5)}$	$A^{(6)}$	$A^{(7)}$	$A^{(8)}$	1A	2A	3A	5A	B*
e_1	1	3	3	3	3	3	3	1	1	1	1	1	1
e_2	1	1	-1	-1	-1	-1	1	1	5	1	-1	0	0
e_3	1	$\sqrt{5}$	1	1	-1	-1	$-\sqrt{5}$	-1	3	-1	0	-b5	*
e_4	1	$-\sqrt{5}$	1	1	-1	-1	$\sqrt{5}$	-1	3	-1	0	*	-b5
e_5	$\begin{pmatrix}1&0\\0&1\end{pmatrix}$	$\begin{pmatrix}-1&1\\1&-1\end{pmatrix}$	$\begin{pmatrix}-1&1\\-3&0\end{pmatrix}$	$\begin{pmatrix}0&-3\\1&-1\end{pmatrix}$	$\begin{pmatrix}3&0\\1&-1\end{pmatrix}$	$\begin{pmatrix}-1&1\\0&3\end{pmatrix}$	$\begin{pmatrix}-1&1\\1&-1\end{pmatrix}$	$\begin{pmatrix}0&-1\\-1&0\end{pmatrix}$	4	0	1	-1	-1

$\left(b5 = \dfrac{-1+\sqrt{5}}{2}\right)$

A $\mathbb{Z}$-basis for $Z(E)$ is given by $z_1 = A^{(1)}$, $z_2 = A^{(2)} + A^{(8)}$, $z_3 = A^{(3)} + A^{(4)} - 2A^{(8)}$, $z_4 = A^{(5)} + A^{(6)} + A^{(8)}$ and $z_5 = A^{(7)} + A^{(8)}$.

<u>Central characters</u> / <u>p-blocks</u>

	z_1	z_2	z_3	z_4	z_5	p = 2	p = 3	p = 5
ω_1	1	4	4	7	4	a	a	a
ω_2	1	2	-4	-1	2	a	b	b
ω_3	1	2b5	4	-3	*	a	c	a
ω_4	1	*	4	-3	2b5	a	d	a
ω_5	1	-1	-1	2	-1	b	b	a

<u>Decomposition numbers</u>

	p = 2		p = 3				p = 5	
	W_1	W_2	W_1	W_2	W_3	W_4	W_1	W_2
V_1	1	0	1	0	0	0	1	0
V_2	1	0	0	1	0	0	0	1
V_3	1	0	0	0	1	0	1	0
V_4	1	0	0	0	0	1	1	0
V_5	0	1	0	2	0	0	2	0

See also exercise 2 for an example with E commutative.

<u>Exercises</u>

1. Use the argument of 2.2.2 to show that if H is a subgroup of G then

$$A(G,\mathrm{Triv}) = i_{H,G}(A(H,\mathrm{Triv})) \oplus \mathrm{Ker}_{A(G,\mathrm{Triv})}(r_{G,H}) .$$

2. Let $G = S_8$, the symmetric group on eight letters, let $H = S_5 \times S_3$ be the subgroup fixing an unordered triple of letters, and let R be the 2-adic integers. Let S be the set of right cosets of H in G, so that S may be thought of as the set of unordered triples from the eight letters.

(i) Show that there are four orbits S_i^2 of G on $S \times S$, with $k_i/|G:H| = 1, 15, 30$ and 10, for $i = 1, 2, 3$ and 4 respectively.

(ii) Using the basis $A^{(1)}, A^{(2)}, A^{(3)}, A^{(4)}$ for E, show that in the regular representation of E, $A^{(2)}$ is represented as the matrix

$$\alpha = \begin{pmatrix} 0 & 15 & & \\ 1 & 6 & 8 & \\ & 4 & 8 & 3 \\ & & 9 & 6 \end{pmatrix} .$$

(iii) Find the eigenvalues of α. Deduce that $A^{(2)}$ generates $\hat{E}$, and E is commutative. Thus the irreducible representations of $\hat{E}$ are the central homomorphisms, and the components of $\overline{R}S$ are the central components.

(iv) Deduce that the representations of $\hat{E}$ are as follows:

	$A^{(1)}$	$A^{(2)}$	$A^{(3)}$	$A^{(4)}$
ω_1	1	15	30	10
ω_2	1	7	-2	-6
ω_3	1	1	-5	3
ω_4	1	-3	3	-1 .

(v) Use 2.6.4(ii) to calculate the dimension of the ordinary representation V_i corresponding to each ω_i.

(vi) Show that $\overline{R}S$ is the direct sum of two indecomposable modules, of dimensions 8 and 48. What is the dimension of the endomorphism ring of each direct summand?

See also [11] for further information.

2.7 Defect Groups

As in section 2.6, we let G act on a set S, and we let $A^{(i)}$ be the standard basis elements of $\overline{E}(S)$ corresponding to $S \times S = \dot{\bigcup_i} S_i^2$. We define a <u>defect group</u> of S_i^2 to be a Sylow p-subgroup of the stabilizer of a point in S_i^2. This is well defined up to conjugacy in G.

2.7.1 Lemma

If D is a p-subgroup of G, then $(\overline{R}S,\overline{R}S)_D^G$ is the linear span in $\overline{E}(S)$ of the $A^{(i)}$ for which D contains a defect group of S_i^2.

Proof

If $(x,y) \in S \times S$, let α be the basis element of $E_{\overline{R}\mathrm{Stab}_D(x,y)}(S)$ corresponding to (x,y), let β be the basis element of $E_{\overline{R}D}(S)$ corresponding to the D-orbit of (x,y), and let S_i^2 be the G-orbit of (x,y). Then

$$\begin{aligned}\mathrm{Tr}_{D,G}(\beta) &= \mathrm{Tr}_{\mathrm{Stab}_D(x,y),G}(\alpha)\\ &= \mathrm{Tr}_{\mathrm{Stab}_G(x,y),G}(|\mathrm{Stab}_G(x,y):\mathrm{Stab}_D(x,y)|\ .\alpha)\\ &= \begin{cases}\text{A non-zero multiple of } A^{(i)} & \text{if } \mathrm{Stab}_D(x,y) \in \mathrm{Syl}_p(\mathrm{Stab}_G(x,y))\\ 0 & \text{otherwise.}\end{cases}\end{aligned}$$

Thus precisely those $A^{(i)}$ for which D contains a defect group of S_i^2 appear as traces of basis elements of $E_{\bar{R}D}(S)$. □

Now suppose e is an idempotent in $\bar{E}(S)$. Then a <u>defect group</u> of e is a minimal subgroup D such that $e \in (\bar{R}S,\bar{R}S)_D^G$. By 2.3.2 and the definition of vertex, if e is primitive, a defect group of e is the same as a vertex of $\bar{R}S.e$. In particular, the defect groups of a primitive idempotent are conjugate in G by 2.5.1.

2.7.2 <u>Proposition</u>

Suppose $e = \Sigma e_i A^{(i)}$ is a primitive idempotent in $\bar{E}(S)$ with defect group D.

(i) D contains a defect group of each S_i^2 such that $e_i \neq 0$.

(ii) There is an i with $e_i \neq 0$ such that D is a defect group of S_i^2.

(iii) Suppose $e \in Z(\bar{E}(S))$ and $\bar{R}$ is a splitting field. Let ω be the corresponding central homomorphism. If $\omega(A^{(i)}) \neq 0$ then D is contained in a defect group of $A^{(i)}$. Thus the defect groups of e are the defect groups of each suborbit for which $e_i \neq 0$ and $\omega(A^{(i)}) \neq 0$ (there are some since $\omega(e) = 1$).

<u>Proof</u>

(i) This follows from 2.7.1 and the definition of defect group .

(ii) Since e is in the sum of the $(\bar{R}S,\bar{R}S)_{D'}^G$ as D' runs over the set of defect groups of the S_i^2 for which $e_i \neq 0$, Rosenberg's lemma (1.5.5) implies that for some such D', $(\bar{R}S,\bar{R}S)_{D'}^G = (\bar{R}S,\bar{R}S)_D^G$. By (i), D contains a conjugate of D', and the result follows by minimality of D.

(iii) $\omega(A^{(i)}e) = \omega(A^{(i)})\omega(e) \neq 0$, so $A^{(i)}e \notin \mathrm{Rad}(e\bar{E})$. But $e\bar{E}$ is a local ring, so $e \in A^{(i)}e\bar{E} \subseteq (\bar{R}S,\bar{R}S)_{D_i}^G$, where D_i is a defect group of $A^{(i)}$. Thus D_i contains a conjugate of D. □

<u>The Classical Case</u>

We have an action of $G \times G$ on $S = \{g \in G\}$ via

$(x,y) : g \to x^{-1}gy$. Then $\mathrm{End}_{\bar{R}(G\times G)}(\bar{R}S) \cong Z(\bar{R}G)$, and the primitive idempotents correspond to the blocks of $\bar{R}G$. Each orbit of $G \times G$ on $S \times S$ (note carefully how $G \times G$ acts on $S \times S$!) contains an element of the form $(1,g)$, and $\mathrm{Stab}_{G\times G}(1,g) = \mathrm{diag}(C_G(g))$. We then say that D is a <u>defect group of the block</u> B of $\bar{R}G$ if $\mathrm{diag}(D)$ is a defect group of the corresponding idempotent in $\mathrm{End}_{\bar{R}(G\times G)}(\bar{R}S)$; namely the vertex of B as a $G \times G$-module.

Thus if we define the <u>defect group</u> of a conjugacy class of elements $g \in G$ to be a Sylow p-subgroup of $C_G(g)$, then by 2.7.2 the defect group of a block B is a maximal defect group over conjugacy classes whose sum is involved in the idempotent e. Moreover, if $\bar{R}$ is a splitting field and ω_e is the central homomorphism corresponding to e, then the minimal defect groups of conjugacy classes of $g \in G$ for which $\omega_e(\sum_{g\in G} g) \neq 0$ are the defect groups of B. If $|D| = p^d$, we say that B is a block of <u>defect</u> d. It turns out that the defect of a block gives some measure of how complicated the representation theory of the block is. Thus as we shall see in 2.7.5 and 2.12.9, a block has defect zero if and only if there is only one indecomposable module in the block, and it has cyclic defect groups if and only if there are only finitely many indecomposable modules in the block.

We refer to the above case as the <u>classical case</u>, since it was the original case investigated by Brauer.

2.7.3 <u>Proposition</u> (Green)

Let B be a block of $\bar{R}G$. Let $P \in \mathrm{Syl}_p(G)$. Then there is an element $g \in G$ such that $P \cap P^g$ is a defect group of B.

<u>Proof</u>

Regard B as an indecomposable trivial source $\bar{R}(G \times G)$-module with vertex $\mathrm{diag}(D)$ as above. Then $P \times P \in \mathrm{Syl}_p(G \times G)$, and so by 2.3.3, $B{\downarrow}_{P\times P}$ has an indecomposable trivial source summand with vertex $\mathrm{diag}(D)$. But by the Mackey decomposition theorem,

$$(\bar{R}G){\downarrow}_{P\times P} = 1_{\mathrm{diag}(G)}{\uparrow}^{G\times G}{\downarrow}_{P\times P}$$

$$= \sum_{\substack{\text{conj. classes} \\ \text{of elts. } g \in G}} 1_{\mathrm{diag}(G)^{(1,g)} \cap (P\times P)}{\uparrow}^{P\times P}$$

$$= \sum 1_{\mathrm{diag}(P \cap P^g)^{(1,g)}}{\uparrow}^{P\times P} .$$

Now every transitive permutation module for a p-group is indecomposable, since Frobenius reciprocity shows that it has a simple socle. Thus D is conjugate to some $P \cap P^g$. □

Remaining in the classical case, we have the following proposition relating defect groups to modules.

2.7.4 Proposition (Green)

(i) Let V be an $\bar{R}G$-module. Let D be a p-subgroup of $H \leq G$, and let e be an idempotent in $(\bar{R}S,\bar{R}S)^{G\times H}_{diag(D)} \subseteq (\bar{R}S,\bar{R}S)^{G\times 1} \cong \bar{R}G$. Then $V{\downarrow_H}.e$ is an $\bar{R}H$-module which is D-projective.

(ii) Suppose B is a block of $\bar{R}G$ with defect group D. Then every indecomposable $\bar{R}G$-module V in B is D-projective.

Proof

(i) Since e is H-invariant, $V{\downarrow_H}.e$ is an $\bar{R}H$-module, which is a direct summand of $V{\downarrow_H}$. Let $X = (\bar{R}S).e$ as a trivial source $\bar{R}(G\times H)$-module. Then X is diag(D)-projective, and so by the Mackey decomposition theorem, $X{\downarrow_{diag(H)}}$ is diag(D)-projective. Write α for the identity endomorphism of X, and write $\alpha = Tr_{diag(D),diag(H)}(\beta)$ with $\beta \in End_{diag(D)}(X)$. Thus for $v \in V$,

$$v = v(e\alpha) = v \sum_g g^{-1}((geg^{-1})\beta)g = \sum_g((vg^{-1})(e\beta))g$$

where g runs over a set of right coset representatives of D in H.

Thus $e\beta$ acts on V as an $\bar{R}D$-module endomorphism and $Tr_{D,H}(e\beta) = 1_V$. The result now follows from 2.3.2.

(ii) This is the case $G = H$ of (i). □

Remark

We shall show in section 2.12 that in fact there is always an indecomposable module in B whose vertex is exactly the defect group D of B.

2.7.5 Corollary (Blocks of Defect 0).

Suppose B is a block of defect 0. Then there is only one indecomposable module in B, and it is both irreducible and projective. B is a complete matrix algebra over a division ring.

Proof

By 2.7.4(ii), $B/J(B)$ is a projective B-module, and so as a B-module, we have $B \cong B/J(B) \oplus J(B)$. Thus $J(B)/J^2(B) = 0$, and so by Nakayama's lemma (1.1.4) $J(B) = 0$. The result now follows from the Wedderburn-Artin structure theorem (1.2.4). □

We shall see in the next section that some questions about blocks may be reduced to questions about blocks of defect 0 (extended first main theorem).

There is also a large body of information available on the structure of blocks with cyclic defect group, see [51].

Exercise

Find the vertices of the summands in exercise 2 of section 2.6.

2.8 The Brauer Homomorphism

Let $D \leq H \leq N_G(D) \leq G$, with D a p-group. Let S^D be the fixed points of D on S. Then S^D is invariant under H, and hence forms a permutation representation of H/D.

We have a natural map

$$Br^D_{H,H} : End_{\bar{R}H}(S) \to End_{\bar{R}H}(S^D) = End_{\bar{R}(H/D)}(S^D)$$

sending a basis element of $End_{\bar{R}H}(S)$ to the same basis element of $End_{\bar{R}H}(S^D)$ if the corresponding H-orbit on $S \times S$ is in $S^D \times S^D$ and to zero otherwise. We define

$$Br^D_{G,H} : End_{\bar{R}G}(S) \to End_{\bar{R}H}(S^D) = End_{\bar{R}(H/D)}(S^D)$$

to be the composite of the inclusion $End_{\bar{R}G}(S) \hookrightarrow End_{\bar{R}H}(S)$ with $Br^D_{H,H}$. Then $Br^D_{G,H}$ is called the Brauer map.

2.8.1 Lemma

$Br^D_{G,H}$ is a ring homomorphism, with kernel $(\bar{R}S,\bar{R}S)^G_{\mathfrak{X}}$, where $\mathfrak{X}$ = {p-subgroups of G not conjugate to a subgroup containing D}.

Proof

$$\Sigma\, c_i A^{(i)} \in Ker(Br^D_{G,H}) \Leftrightarrow S_i^2 \cap (S^D \times S^D) = \emptyset \text{ whenever } c_i \neq 0$$

$\Leftrightarrow D$ is not conjugate to a subgroup of a defect group of S_i^2 whenever $c_i \neq 0$

$\Leftrightarrow \Sigma\, c_i A^{(i)} \in (\bar{R}S,\bar{R}S)^G_{\mathfrak{X}}$ by 2.7.1.

In particular, if we regard $End_{\bar{R}H}(\bar{R}(S^D))$ as a subring of $End_{\bar{R}H}(\bar{R}S)$, we have

$$End_{\bar{R}H}(\bar{R}S) = (\bar{R}S,\bar{R}S)^H_{\mathfrak{X}} \oplus End_{\bar{R}H}(\bar{R}(S^D))$$

as vector spaces, and the map $Br^D_{H,H}$ is the projection onto the second factor, and is a homomorphism, since $(\bar{R}S,\bar{R}S)^H_{\mathfrak{X}}$ is an ideal. □

Returning to the classical case, suppose $C_G(D) \leq K \leq N_G(D)$. Then $S^{diag(D)}$ is the set of elements of $C_G(D)$, and we have an inclusion $End_{\bar{R}diag(N_G(D))}(\bar{R}(S^{diag(D)})) \subseteq End_{\bar{R}diag(C_G(D))}(\bar{R}(S^{diag(D)})) \cong Z(\bar{R}C_G(D)) \subseteq Z(\bar{R}K)$. Composing this with $Br^{diag(D)}_{GxG,diag(N_G(D))}$ gives a homomorphism $br^D_{G,K} : Z(\bar{R}G) \to Z(\bar{R}K)$, which is also called the <u>Brauer map</u>. This is the map sending a class sum to the sum of those elements of the class lying in $C_G(D)$. By 2.8.1, $br^D_{G,K}$ is a ring homomorphism whose kernel is the ideal $(\bar{R}G,\bar{R}G)^{G\times G}_{\mathfrak{X}} \subseteq (\bar{R}G,\bar{R}G)^{G\times G} \cong Z(\bar{R}G)$ with $\mathfrak{X}$ = {p-subgroups of $G \times G$ not conjugate to a subgroup containing diag(D)}. Moreover by 2.7.1, this is the linear span in $Z(\bar{R}G)$ of those conjugacy class sums for elements $g \in G$ for which a Sylow p-subgroup of $C_G(g)$ does not contain a conjugate of D.

<u>Notation</u>

If $\mathfrak{X}$ is a collection of subgroups of G, we denote by $Z_{\mathfrak{X}}(\bar{R}G)$ the subspace of $Z(\bar{R}G)$ spanned by those class sums for conjugacy classes with a defect group contained in an element of $\mathfrak{X}$.

We have thus proved the following theorem.

2.8.2 <u>Theorem</u>

Let D be a p-subgroup of G, and let $C_G(D) \le K \le N_G(D)$. Then the map $br^D_{G,K} : Z(\bar{R}G) \to Z(\bar{R}K)$ given by sending each class sum to the sum of these elements lying in $C_G(D)$, is a ring homomorphism with kernel $Z_{\mathfrak{X}}(\bar{R}G)$, where $\mathfrak{X}$ is the set of p-subgroups of G not conjugate to a subgroup containing D. □

Now let $1 = e_1 + \ldots + e_s$ be the idempotent decomposition in $Z(\bar{R}G)$ corresponding to the block decomposition $\bar{R}G = B_1 \oplus .. \oplus B_s$. Suppose e is a primitive idempotent in $Z(\bar{R}K)$. Then

$$e = e.br^D_{G,K}(1) = e.br^D_{G,K}(e_1) + .. + e.br^D_{G,K}(e_s).$$

Since e is primitive, there is one and only one i with $e = e.br^D_{G,K}(e_i)$, and $e.br^D_{G,K}(e_j) = 0$ for $j \neq i$. If e corresponds to the block b of $\bar{R}K$, we write $b^G = B_i$, and we say B_i is the <u>Brauer correspondent</u> of b. If $\bar{R}$ is a splitting field, then we may reformulate this in terms of central homomorphisms as follows. If ω is the central homomorphism corresponding to b, then $br^D_{G,K}.\omega : Z(\bar{R}G) \to \bar{R}$ is a central homomorphism, and b^G is the block of G corresponding to it.

We now prove the classical and permutation versions of Brauer's first main theorem.

2.8.3 <u>Lemma</u>

Let $N = N_G(D)$. If the G-orbit S^2_i has defect group D, then

$S_1^2 \cap (S^D \times S^D)$ is a single N-orbit with defect group D. Each N-orbit on $S^D \times S^D$ with defect group D is of this form.

Proof

Let (x,y) and (x',y') be elements of $S_1^2 \cap (S^D \times S^D)$, and choose $g \in G$ with $(x,y)g = (x',y')$. Then D and D^g are Sylow p-subgroups of $\mathrm{Stab}_G(x',y')$ and so we may choose an element $h \in \mathrm{Stab}_G(x',y')$ with $D^h = D^g$. Thus $gh^{-1} \in N_G(D)$, and $(x',y') = (x,y)gh^{-1}$.

Conversely, if (x,y) is in an N-orbit on $S^D \times S^D$ with defect group D, then $D \in \mathrm{Syl}_p(\mathrm{Stab}_N(x,y))$. If $D < D_1 \in \mathrm{Syl}_p(\mathrm{Stab}_G(x,y))$ then $D < N_{D_1}(D) \le N \cap \mathrm{Stab}_G(x,y) = \mathrm{Stab}_N(x,y)$. This contradiction proves the last statement. □

2.8.4 Proposition

Let $N = N_G(D)$. Then $\mathrm{Br}^D_{G,N}$ induces an isomorphism

$$(\bar{R}S,\bar{R}S)^G_D \Big/ \sum_{D'<D} (\bar{R}S,\bar{R}S)^G_{D'} \cong (\bar{R}S^D,\bar{R}S^D)^N_D \cong (\bar{R}S^D,\bar{R}S^D)^{N/D}_1$$

Proof

By 2.8.1 and 2.3.1(vi), the kernel of $\mathrm{Br}^D_{G,N}$ on $(\bar{R}S,\bar{R}S)^G_D$ is $\sum_{D'<D} (\bar{R}S,\bar{R}S)^G_{D'}$. By 2.8.3, the image is exactly $(\bar{R}S^D,\bar{R}S^D)^N_D$. (Notice that D is contained in every defect group for $\mathrm{End}_{\bar{R}N}(\bar{R}S^D)$). □

2.8.5 Brauer's First Main Theorem (permutation version)

$\mathrm{Br}^D_{G,N}$ establishes a one-one correspondence between equivalence classes of primitive idempotents of $(\bar{R}S,\bar{R}S)^G$ with defect group D (recall that two primitive idempotents are equivalent if they lie in the same Wedderburn component of $(\bar{R}S,\bar{R}S)^G/J((\bar{R}S,\bar{R}S)^G)$) and equivalence classes of primitive idempotents of $(\bar{R}S^D,\bar{R}S^D)^N$ with defect group D (or equivalently with equivalence classes of primitive idempotents of $(\bar{R}S^D,\bar{R}S^D)^{N/D}_1$). In particular if $(\bar{R}S,\bar{R}S)^G_D$ is commutative, then $\mathrm{Br}^D_{G,N}$ establishes a one-one correspondence between primitive idempotents of $(\bar{R}S,\bar{R}S)^G$ with defect group D and primitive idempotents of $(\bar{R}S^D,\bar{R}S^D)^N$ with defect group D.

Proof

This follows immediately from 2.8.4 and the idempotent refinement theorem (1.5.1). □

2.8.6 Brauer's First Main Theorem (classical version)

Let $N = N_G(D)$. Then $b \to b^G$ gives a one-one correspondence between blocks of $\bar{R}N$ with defect group D and blocks of $\bar{R}G$ with defect group D.

Proof

This follows immediately by applying 2.8.5 to the classical case. □

Warning

The blocks of $\bar{R}N$ with defect group D are not in general in one-one correspondence with blocks of $\bar{R}(N/D)$ of defect zero.

To reduce the case of blocks of defect zero, we have the following extension of 2.8.6, whose proof we shall omit (see [65]).

2.8.6a The Extended First Main Theorem

The following are in natural one-one correspondence.

(i) Blocks of G with defect group D.

(ii) Blocks of $N_G(D)$ with defect group D.

(iii) $N_G(D)$-conjugacy classes of blocks of $C_G(D)$ with D as defect group in $N_G(D)$, (here, we have $N_G(D) \times N_G(D)$ acting on the set of elements of $C_G(D)$).

(iv) (assuming $\bar{R}$ is a splitting field for $\bar{R}C_G(D)$), $N_G(D)$-conjugacy classes of blocks b of $C_G(D)$ with D as defect group in $DC_G(D)$ and $|N_G(b):DC_G(D)|$ coprime to p.

(v) (assuming $\bar{R}$ is a splitting field for $\bar{R}C_G(D)$), $N_G(D)$-conjugacy classes of blocks b of defect zero of $DC_G(D)/D$ with $|N_G(b):DC_G(D)|$ coprime to p. □

Examples

Let G be a simple group of Lie type (see [28]) in characteristic p. Then G has two p-blocks, namely the principal block (i.e. the block with the trivial representation in it) and a block of defect zero consisting of the Steinberg representation, whose degree is equal to the order of the Sylow p-subgroup.

Now let M denote the Monster simple group (sometimes denoted F_1). Then there is an elementary abelian subgroup D of order four whose normalizer has shape $2^2.{}^2E_6(2).S_3$. From the above, we know that in characteristic two, $C_G(D)/D$ has exactly one block b of defect zero, and that therefore $N_G(D) = N_G(b)$. Theorem 2.8.6a now tells us that M has a single block with defect group D.

The following is Nagao's module theoretic version of Brauer's second main theorem.

2.8.7 Theorem (Nagao)

Let e be a central idempotent in $\bar{R}G$, let D be a p-subgroup of G, and let $C_G(D) \le K \le N_G(D)$. Let $\mathcal{H} = \{\text{p-subgroups } Q \le K: Q \not\ge D\}$. If $V.e = V$ then $V{\downarrow_K} - V{\downarrow_K}.\mathrm{br}^D_{G.K}(e) \in a(K, \mathcal{H})$.

Proof

Embed $\bar{R}K$ in $(\bar{R}S,\bar{R}S)^{G\times K}$ in the obvious way, and let $f = e - br^{D}_{G,K}(e)$ as an element of $(\bar{R}S,\bar{R}S)^{G\times K}$. Then by 2.8.2, $f \in (\bar{R}S,\bar{R}S)^{G\times K}_{diag(\mathcal{H})}$. Let $f = \Sigma f_i$ be a decomposition of f as a sum of primitive orthogonal idempotents in $(\bar{R}S,\bar{R}S)^{G\times K}_{diag(\mathcal{H})}$. Then by Rosenberg's lemma (1.5.5), each f_i is in $(\bar{R}S,\bar{R}S)^{G\times K}_{diag(Q)}$ for some $Q \in \mathcal{H}$, and so by 2.7.4(i), $V{\downarrow}_H.f_i$ is Q-projective. Thus

$$V{\downarrow}_H = V{\downarrow}_H.e = V{\downarrow}_H.br^{D}_{G,K}(e) \oplus V{\downarrow}_H.f$$
$$= V{\downarrow}_H.br^{D}_{G,K}(e) \oplus (\bigoplus_i V{\downarrow}_H.f_i)$$

and the theorem is proved. □

2.9 Origins of Species

Before we define the origins of a species, we need an integrality theorem.

2.9.1 Theorem

Let $H \le G$. Then $A(H)$ is integral as an extension of $Im(r_{G,H})$. (c.f. 2.2.2(ii)).

Proof

If $\alpha \in A(H)$ then α has only finitely many images $\alpha_1, .., \alpha_r$ under the action of $N_G(H)$. Thus α satisfies $(x-\alpha_1)(x-\alpha_2) .. (x-\alpha_r) = 0$, and so α is integral over $A(H)^{N(H)}$, the fixed points of $N_G(H)$ on $A(H)$. Thus we only need show that $A(H)^{N(H)}$ is integral over $Im(r_{G,H})$.

Let $\alpha \in A(H)^{N(H)}$. For any $K < H$ we set

$$X_K = \{r_{H^g,K}(\alpha^g) : K < H^g\} .$$

Denote by U_K the subring of $A(K)$ generated by $Im(r_{G,K})$ and X_K. We may assume inductively that U_K is finitely generated as a module for $Im(r_{G,K})$. We claim that

$$Im(r_{G,H}) + \sum_{K<H} i_{K,H}(U_K)$$

is a subring of $A(H)$, finitely generated as a module for $Im(r_{G,H})$, and containing the element a.

(i) We first show that it is a ring. It is clear that $\mathrm{Im}(r_{G,H}).i_{K,H}(U_K) \subseteq i_{K,H}(U_K)$ by 2.1.2(iv). If $K < H$ and $L < H$, then

$$i_{K,H}(U_K).i_{L,H}(U_L) \subseteq \sum_{g \in H} i_{K\cap L^g,H}(r_{K,K\cap L^g}(U_K).r_{L^g,K\cap L^g}(U_{L^g}))$$

by the Mackey decomposition theorem. But

$$r_{K,K\cap L^g}(U_K) \subseteq \langle \mathrm{Im}(r_{G,K\cap L^g}), r_{K,K\cap L^g}(X_K)\rangle \subseteq U_{K\cap L^g} .$$

(ii) It is finitely generated as a module for $\mathrm{Im}(r_{G,H})$. By induction, for any $K < H$, we have

$$U_K = \sum_{b \in Y_K} \mathrm{Im}(r_{G,K}).b$$

with Y_K a finite set. So

$$i_{K,H}(U_K) = \sum_{b \in Y_K} \mathrm{Im}(r_{G,H}).i_{K,H}(b).$$

(iii) It contains α.
Since α is invariant under $N_G(H)$, we have

$$\alpha\uparrow^G\downarrow_H = |N_G(H):H|\alpha + \sum_{\substack{HgH \\ \text{s.t.} H\cap H^g < H}} \alpha^g\downarrow_{H\cap H^g}\uparrow^H . \quad \square$$

2.9.2 Proposition

Let s be a species of $A(G)$. The following conditions on a subgroup H are equivalent.

(i) $\mathrm{Ker}(s) \geq \mathrm{Ker}(r_{G,H})$

(ii) $\mathrm{Ker}(s) \not\geq \mathrm{Im}(i_{H,G})$

(iii) There is a species t of $A(H)$ such that for all $x \in A(G)$ $(s,x) = (t,x\downarrow_H)$.

Proof

(i) $\Leftrightarrow$ (ii) by 2.2.2

(ii) $\Leftrightarrow$ (iii) by 2.9.1 and 1.8.1. $\square$

Note that in 2.9.2 (iii) the species t need not be unique. We write $t \sim s$, and say t fuses to s if (iii) is satisfied.

We say s factors through H if the equivalent conditions of 2.9.2 are satisfied. An origin of s is a subgroup minimal among those through which s factors. Thus if H is an origin of s and $K < H$ then s vanishes on all modules induced from K.

2.9.3 Proposition

Let s be a species of $A(G)$. Then the origins of s form a single conjugacy class of subgroups.

Proof

Let H_1 and H_2 be two origins of s. Then since Ker(s) is a prime ideal,

$$\mathrm{Ker}(s) \nsupseteq \mathrm{Im}(i_{H_1,G}).\mathrm{Im}(i_{H_2,G}) \leq \sum_{x \in G} \mathrm{Im}(i_{H_1 \cap H_2^x, G}) .$$

So for some $x \in G$, $\mathrm{Ker}(s) \nsupseteq \mathrm{Im}(i_{H_1 \cap H_2^x})$. Hence by minimality $H_1 = H_2^x$. □

In section 2.14 we shall clarify the structures of origins and their relationships to vertices of the species.

2.10 The Induction Formula

Let s be a species of A(G) with origin H, and let V be a module for a subgroup K of G. We want a formula for $(s, V{\uparrow}^G)$ in terms of the species of K.

Let t be a species of H fusing to s. Then

$$(s,V{\uparrow}^G) = (t,V{\uparrow}^G{\downarrow}_H)$$

$$= \sum_{HgK} (t,\ V^g{\downarrow}_{H \cap K^g}{\uparrow}^H) \text{ by the Mackey theorem.}$$

Now if $H \nsubseteq K^g$, then $(t,V^g{\downarrow}_{H \cap K^g}{\uparrow}^H)$ is zero by 2.9.2, since H is an origin of t. Thus we have the following formula.

(1) $$(s,V{\uparrow}^G) = \sum_{H^g \leq K} |N_G(H^g):N_K(H^g)|\ (t^g,V{\downarrow}_{H^g})$$

The sum runs over K-conjugacy classes of G-conjugates of H contained in K.

In order to convert this into a formula involving species of K, we must examine the number of species of K fusing to s.

2.10.1 Theorem

Let s be a species of A(G) with origin H. Regard s as a species of $\mathrm{Im}(r_{G,H})$. Then s extends uniquely to a species t of $A(H)^{N(H)}$, and $N_G(H)$ is transitive on the extensions $t_1, \ldots, t_r$ of t to a species of A(H). The number of extensions is $r = |N_G(H) : \mathrm{Stab}_G(t_1)|$.

Proof

By 2.9.2(iii), s certainly extends to a species of $A(H)^{N(H)}$ and a species of A(H). Let t be an extension of s to $A(H)^{N(H)}$.

Then for $x \in A(H)^{N(H)}$, the Mackey theorem and the fact that H is an origin for s imply that $(t, x\uparrow^G\downarrow_H) = |N_G(H):H|(t,x)$. Thus $(t,x) = (s, x\uparrow^G)/|N_G(H):H|$ is uniquely determined by s.

Now suppose that t_1 and t_2 are two extensions of t to $A(H)$, and that $t_1 \neq t_2^g$ for all $g \in N_G(H)$. Then by 2.2.1 there is an element $x \in A(H)$ such that $(t_1,x) = 0$ and $(t_2,x^g) = (t_2^{g^{-1}},x) = 1$ for all $g \in N_G(H)$. Let $y = \prod_{g \in N(H)} x^g$. Then $0 = (t_1,y) = (t,y) = \prod_{g \in N(H)} (t_2,x^g) = 1$. This contradiction proves the theorem. The formula for the number of extensions is clear. □

By 2.10.1, the contribution in (1) from a particular conjugate H^g is

$$\sum_{t^g \sim s} |\mathrm{Stab}_G(t^g):N_K(H^g)|(t^g, V\downarrow_{H^g}) .$$

In this expression, t^g runs over the species of H^g fusing to s. If s_o is a species for K fusing to s, and with origin H^g, then by 2.10.1, the number of t^g fusing to s_o is

$$|N_K(H^g):\mathrm{Stab}_K(t^g)| = |N_G(H^g) \cap \mathrm{Stab}_G(s_o):\mathrm{Stab}_G(t^g)| .$$

Thus

$$\sum_{t^g \sim s_o} |\mathrm{Stab}_G(t^g):N_K(H^g)|(t^g,V\downarrow_{H^g}) = |N_G(H^g) \cap \mathrm{Stab}_G(s_o):N_K(H^g)|(s_o,V)$$

Hence we can rewrite (1) as follows.

$$(2.10.2) \quad (s,V\uparrow^G) = \sum_{s_o \sim s} |N_G(\mathrm{Orig}(s_o)) \cap \mathrm{Stab}_G(s_o):N_K(\mathrm{Orig}(s_o))|(s_o,V).$$

In this expression, s_o runs over the species of K fusing to s, and $\mathrm{Orig}(s_o)$ is any origin of s_o.

The expression 2.10.2 is called the induction formula, and it is a generalization of the usual formula for an induced character.

2.11 Brauer Species

A species s of $A_k(G)$ is called a Brauer species if its origins have order coprime to p. By Maschke's theorem, the Brauer species vanish on $A_o(G,1)$, and may thus be thought of as species of $A(G)/A_o(G,1)$. We shall first construct some Brauer species, and then show that we have constructed them all. It will turn out that if k is a splitting field, then there are as many Brauer species as there are p-regular conjugacy classes of G.

Let $\hat{k}$ be the algebraic closure of k, and let γ be the p'-part of the exponent of G. Then the γth roots of unity in $\hat{k}$ and in $\mathbb{C}$ both form a cyclic group of order γ. Choose an isomorphism between these cyclic groups. Let g be a p'-element of G. Given an kG-module V, we restrict it to $\langle g\rangle$ and extend the field to $\hat{k}$. Then each eigenvalue of g is a γth root of unity in $\hat{k}$, and we define (b_g,V) to be the sum of the corresponding roots of unity in $\mathbb{C}$. It is clear that b_g is a Brauer species with $\langle g\rangle$ as an origin.

2.11.1 <u>Lemma</u>

Let b_g be as above. Then there is an element $y \in A(G,1)$ with $(b_g,y) \neq 0$ and $(b_{g'},y) = 0$ for every $b_{g'} \neq b_g$.

<u>Proof</u>

This is clear if $G = \langle g\rangle$, since $A(\langle g\rangle) = A(\langle g\rangle,1)$ and the b_g are linearly independent (2.2.1). Let $x \in A(\langle g\rangle,1)$ with this property, and let $y = x\uparrow^G$. Then the induction formula 2.10.2 (which is much easier for the b_g than for the general species) shows that $(b_g,y) \neq 0$ and $(b_{g'},y) = 0$ whenever $b_{g'} \neq b_g$. □

2.11.2 <u>Proposition</u>

If $(b_g,W_1) = (b_g,W_2)$ for all p'-elements g then W_1 and W_2 have the same composition factors.

<u>Proof</u>

We may replace W_1 and W_2 by completely reducible representations with the same composition factors, without affecting the values of (b_g,W_1) and (b_g,W_2). Let the irreducible kG-modules be $V_1, \ldots, V_r$ and let the multiplicity of V_i in W_1 be a_i and in W_2 be b_i. By the Wedderburn-Artin structure theorem (1.2.4), we may choose elements $x_i \in kG$ with trace δ_{ij} on V_j.

Since the trace of an element of G is equal to the trace of its p'-part, the hypothesis tells us that every element of kG has the same trace on W_1 as on W_2. In particular, the elements x_i do, and so $a_i \equiv b_i \mod p$. Thus we may strip off some common direct summands, divide every multiplicity by p, and start again. The result now follows by induction. □

2.11.3 <u>Theorem</u>

$A(G) = A(G,1) \oplus A_0(G,1)$, and $A(G,1)$ is semisimple. The following are equivalent.

(i) s is a Brauer species

(ii) s vanishes on $A_o(G,1)$

(iii) s is of the form b_g for some p'-element $g \in G$.

Proof

By 2.11.2, the number of different b_g is at least $\dim(A(G)/A_o(G,1))$. But each species b_g is a species of $A(G)/A_o(G,1)$, and by 2.2.1 they are all linearly independent. Thus we have equality, and it follows that $A(G)/A_o(G,1)$ is semisimple, and its species are precisely the b_g. This proves the equivalence of (i), (ii) and (iii).

Now consider the Cartan homomorphism

$$c : A(G,1) \hookrightarrow A(G) \twoheadrightarrow A(G)/A_o(G,1).$$

By 2.11.1, this is surjective. By the arguments of section 1.5, $\dim(A(G,1)) = \dim(A(G)/A_o(G,1))$, and so c is an isomorphism. Letting $e = c^{-1}(1)$, we have $A(G,1) = e.A(G)$ and $A_o(G,1) = (1-e)A(G)$. □

Note that if k is a splitting field then the number of different b_g is equal to the number of p-regular conjugacy classes of G, by an argument similar to 2.11.1. Thus in this case, the number of p-regular conjugacy classes is equal to the number of irreducible modules.

Exercises

1. Suppose $(\hat{R},R,\bar{R})$ is a splitting p-modular system for G. Let X denote the ordinary character table of $\hat{R}G$-modules, with the columns corresponding to p-singular elements (i.e. elements of order divisible by p) deleted. Let D denote the decomposition matrix, and C the Cartan matrix. Denote by T the Brauer character table of irreducible modules (i.e. the table whose columns are labelled by the p-regular conjugacy classes, rows are labelled by the irreducible modules, and entries (b_g,V)) and U the Brauer character table of projective indecomposable modules. Show that the following relations hold.

$$X = DT; \; U = D^tX; \quad C = D^tD; \quad U = CT.$$

(see section 1.7)

These are sometimes called the modular orthogonality relations. We shall introduce a generalized form of these relations in section 2.21.

2. Write down the ordinary character table of A_5. Find the central homomorphism associated with each ordinary character, and hence find the blocks of A_5 in characteristic two. What are the defect groups

of the blocks? Using the isomorphism $A_5 \cong L_2(4)$, show that there are two isomorphism classes of two-dimensional irreducible modules over a large enough field $\bar{R}$ of characteristic two. Write down the decomposition matrix, Cartan matrix and Brauer character tables of irreducible and projective indecomposable modules.

Denote the simple $\bar{R}G$-modules by I, 2, 2' and 4 (the numbers refer to the dimensions). Use the action of A_5 on the cosets of a Sylow 5-normalizer to construct a module whose structure is

$$\begin{array}{c} I \\ 2 \oplus 2' \\ I \end{array}$$

(i.e. the socle has dimension one, and is contained in the radical which has codimension one, the quotient being isomorphic to the direct sum of 2 and 2') Hint: use the results of section 2.6.

Show that $\dim \mathrm{Ext}^1_G(I,I) = 0$, using the fact that A_5 has no subgroup of index two. Deduce that P_1 has structure as follows.

$$\begin{array}{ccc} & I & \\ 2 & & 2' \\ I & \oplus & I \\ 2' & & 2 \\ & I & \end{array}$$

Find the structures of the remaining projective indecomposables.

3. Repeat exercise two for $L_3(2)$, and for any other groups that take your fancy. Some large examples are worked out in [11] and [12]; see also the appendix.

Remark

One of the most difficult problems in modular representation theory is to find the decomposition matrices for particular groups modulo particular primes. This problem has not even been solved in general for the symmetric groups (although Lusztig's conjecture in characteristic p, if proved, would give an answer in terms of the so-called Kazhdan-Lusztig polynomials), despite the fact that so much is known about the ordinary representation theory. A remarkable fact about the representation theory of the symmetric groups is that every field is a splitting field!

2.12 Green Correspondence and the Burry-Carlson Theorem

For this section we assume our ring $\Gamma \in \{R,\bar{R}\}$.

Let D be a fixed p-subgroup of G and let H be a subgroup of G containing $N_G(D)$. We shall investigate the modules with vertex D, by means of restriction and induction between H and G. Our main tool is, of course, the Mackey decomposition theorem.

Let

$$\mathfrak{X} = \{X \le G : X \le D^g \cap D \quad \text{for some} \quad g \in G\setminus H\}$$

$$\mathfrak{Y} = \{Y \le G : Y \le D^g \cap H \quad \text{for some} \quad g \in G\setminus H\} .$$

Note that $\mathfrak{X} \subseteq \mathfrak{Y}$ and $D \notin \mathfrak{Y}$.

2.12.1 Lemma

Let W be an indecomposable D-projective ΓH-module.

(i) Let $W{\uparrow}^G{\downarrow}_H \cong W \oplus W'$. Then $W' \in a(H, \mathfrak{Y})$.

(ii) Let $W{\uparrow}^G \cong V \oplus V'$ with W a summand of $V{\downarrow}_H$ and V indecomposable. Then $V' \in a(G, \mathfrak{X})$.

Proof

(i) Let U be an indecomposable D-module with $U{\uparrow}^H \cong W \oplus W_o$. Then

$$U{\uparrow}^G{\downarrow}_H \cong W{\uparrow}^G{\downarrow}_H \oplus W_o{\uparrow}^G{\downarrow}_H .$$

But by the Mackey decomposition theorem,

$$U{\uparrow}^G{\downarrow}_H \cong U{\uparrow}^H \oplus U' \quad \text{with} \quad U' \in a(H, \mathfrak{Y}).$$

Thus

$$W{\uparrow}^G{\downarrow}_H \oplus W_o{\uparrow}^G{\downarrow}_H \cong W \oplus W_o \oplus U'$$

and so by the Krull-Schmidt theorem, $W{\uparrow}^G{\downarrow}_H \cong W \oplus W'$ with $W' \in a(H, \mathfrak{Y})$.

(ii) V' is D-projective. Suppose V' is not $\mathfrak{X}$-projective. Choosen an indecomposable summand V_1 of V' which is not $\mathfrak{X}$-projective, and suppose $D_1 \le D$ is a vertex of V_1. Let U_1 be a source of V_1. Then U_1 is a summand of $V_1{\downarrow}_{D_1}$, and so for some indecomposable summand W_1 of $V_1{\downarrow}_H$, U_1 is a summand of $W_1{\downarrow}_{D_1}$. Thus W_1 is not $\mathfrak{Y}$-projective, and hence $V'{\downarrow}_H$ is not $\mathfrak{Y}$-projective, contradicting (i). □

2.12.2 Theorem (Green Correspondence)

There is a one-one correspondence between indecomposable ΓG-modules with vertex D and indecomposable ΓH-modules with vertex D given as follows.

(i) If V is an indecomposable ΓG-module with vertex D, then $V{\downarrow}_H$ has a unique indecomposable summand $f(V)$ with vertex D, and $V{\downarrow}_H - f(V) \in a(H,\mathfrak{Y})$.

(ii) If W is an indecomposable ΓH-module with vertex D, then $W{\uparrow}^G$ has a unique indecomposable summand $g(W)$ with vertex D, and $W{\uparrow}^G - g(W) \in a(G,\mathfrak{X})$.

(iii) $f(g(W)) = W$ and $g(f(V)) = V$.

(iv) f and g take trivial source modules to trivial source modules.

Proof

(i) Let S be a source of V and let $S{\uparrow}^H \cong W \oplus W'$ with W an indecomposable module such that V is a summand of $W{\uparrow}^G$. By lemma 2.12.1(i), W is the only summand of $W{\uparrow}^G{\downarrow}_H$ with D as vertex, and the rest lie in $a(H,\mathfrak{Y})$. But some summand of $V{\downarrow}_H$ has vertex D, since V is a summand of $V{\downarrow}_H{\uparrow}^G$, and so we take $W = f(V)$.

(ii) Choose an indecomposable summand V of $W{\uparrow}^G$ such that W is a summand of $V{\downarrow}_H$. Then by 2.12.1(ii), $W{\uparrow}^G \cong V \oplus V'$ with $V' \in a(G,\mathfrak{X})$. We take $V = g(W)$.

(iii) and (iv) are clear from (i) and (ii). □

The following remarkable theorem gives us more information about induction and restriction in this situation.

2.12.3 Theorem (D. Burry and J. Carlson)

Let V be an indecomposable ΓG-module such that $V{\downarrow}_H$ has a direct summand W with vertex D. Then V has vertex D, and V is the Green correspondent $g(W)$.

Proof

Let $e = \mathrm{Tr}_{D,H}(\alpha) \in (V,V)^H_D$ be the idempotent corresponding to the summand W of $V{\downarrow}_H$. By 2.3.1(vi), we have

$$\begin{aligned}\mathrm{Tr}_{D,G}(\alpha) &= \sum_{DgH} \mathrm{Tr}_{D^g\cap H,H}(\alpha g) \\ &= e + \sum_{\substack{DgH \\ g\notin H}} \mathrm{Tr}_{D^g\cap H,H}(\alpha g) \\ &\equiv e \mod (V,V)^H_{\mathfrak{Y}} .\end{aligned}$$

Since W is not $\mathfrak{Y}$-projective, $e \notin (V,V)^H_{\mathfrak{Y}}$, and so $Tr_{D,G}(\alpha)$ is an idempotent in $(V,V)^G/((V,V)^G \cap (V,V)^H_{\mathfrak{Y}})$. Since $(V,V)^G$ is a local ring, this means $(V,V)^G = (V,V)^G.Tr_{D,G}(\alpha) \subseteq (V,V)^G_D$, and so V is D-projective. Hence V has vertex D and W is its Green correspondent. □

We shall now reinterpret the Green correspondence in terms of the structure of $A(G)$. Recall that $A(G,H)$ is the ideal of $A(G)$ spanned by the H-projective modules, $A'(G,H)$ is the ideal spanned by the $A(G,K)$ for all $K < H$, and $A_0(G,H)$ is the ideal spanned by elements of the form $X - X' - X''$ where $0 \to X' \to X \to X'' \to 0$ is a short exact sequence of ΓG-modules which splits on restriction to H.

2.12.4 <u>Lemma</u>

If $H \trianglelefteq G$ then

$$A(G) = A(G,H) \oplus A_0(G,H).$$

The idempotent generators of $A(G,H)$ and $A_0(G,H)$ lie in $A(G,Triv)$.

<u>Proof</u>

By 2.11.3,

$$A(G/H) = A(G/H,1) \oplus A_0(G/H,1).$$

Identifying $A(G/H)$ with its image under the natural inclusion $A(G/H) \hookrightarrow A(G)$, we have

$$A(G) = A(G).A(G/H) = A(G).A(G/H,1) + A(G).A_0(G/H,1)$$
$$= A(G,H) + A_0(G,H).$$

Since clearly $A(G,H).A_0(G,H) = 0$, this proves the direct sum decomposition. Since $A(G/H,1) \subseteq A(G,Triv)$, the idempotent generators are in $A(G,Triv)$. □

2.12.5 <u>Lemma</u>

Suppose H is a subgroup of G containing the normalizer of the p-subgroup D. Then $r_{G,H}$ and $i_{H,G}$ induce inverse isomorphisms

$$a(G,D)/a'(G,D) \cong a(H,D)/a'(H,D).$$

These isomorphisms send trivial source modules to trivial source modules.

<u>Proof</u>

This is clear from 2.12.2. □

2.12.6 <u>Theorem</u> (Conlon)

(i) $A(G,H)$ is a direct summand of $A(G)$, whose idempotent generator lies in $A(G,Triv)$.

(ii) $A(G,H)$ has a canonical direct summand $A''(G,H)$ with

$$A(G,H) = A'(G,H) \oplus A''(G,H).$$

(iii) $A(G,H) = \bigoplus_{D \leq H} A''(G,D)$, where D runs over one representative of each G-conjugacy class of p-subgroups of G contained in H.

Proof

We prove these results by induction on $|H|$. By 2.11.3, they are true for $H = 1$, since $A'(G,1) = 0$.

Suppose the theorem is true for all $K < H$. Let e_K be the idempotent generator for $A(G,K)$. Then $e_H' = 1 - \prod_{K<H} (1-e_K)$ is the idempotent generator for $A'(G,H)$, so that e_H' lies in $A(G,Triv)$. Put

$$A''(G,H) = A(G,H).(1-e_H').$$

Then by 2.12.5,

$$A(N_G(H),H)/A'(N_G(H),H) \cong A(G,H)/A'(G,H) \cong A''(G,H)$$

is an isomorphism sending trivial source modules to trivial source modules. In particular, by 2.12.4, $A''(G,H)$ has an idempotent generator lying in $A(G,Triv)$. Thus (i) and (ii) are proved. (iii) follows since each $A''(G,D)$ has a basis consisting of modules with vertex D, modulo $A'(G,D)$. □

We now have a theorem relating the Green correspondence to the Brauer homomorphism.

2.12.7 Theorem

Let D be a p-subgroup of G, and let $N = N_G(D)$. Denote by f and g the Green correspondence between modules for G and N with vertex D, as in 2.12.2. If V is an indecomposable $\overline{R}G$-module with vertex D, and e is the block idempotent for $\overline{R}G$ with $V = V.e$, then $f(V) = f(V).br^D_{G,N}(e)$.

Proof

By 2.8.7, $V{\downarrow}_N - V{\downarrow}_N.br^D_{G,N}(e)$ does not have $f(V)$ as a summand. The result thus follows from 2.12.2. □

This together with the following theorem shows how to reduce questions about the representation theory of a block to questions about representations of the defect group.

2.12.8 Theorem

Let D be a normal p-subgroup of G, and let B be a block of $\overline{R}G$ with defect group D and block idempotent e. Then every indecomposable $\bar{R}D$-module that is not induced from a proper subgroup is the source of some indecomposable module in B.

Proof

The only simple $\bar{R}D$-module is the trivial module 1_D. So if $1_D\uparrow^G.e = 0$ then it would follow that for every $\bar{R}D$-module V we have $V\uparrow^G.e = 0$. Since $\bar{R}D\uparrow^G.e = B$, this is not the case. So $1_D\uparrow^G.e \neq 0$, and for every $\bar{R}D$-module V we have $V\uparrow^G.e \neq 0$. Since $V\uparrow^G\downarrow_D$ is a direct sum of conjugates of V, it follows that if V has vertex D then so does every indecomposable summand of $V\uparrow^G.e$.

2.12.9 Corollary (representation types of blocks)

A block B of $\bar{R}G$ has finite representation type (i.e. there are only finitely many isomorphism classes of indecomposable modules in B) if and only if a defect group D of B is cyclic.

Proof

By 2.12.7 and Brauer's first main theorem (2.8.6) it suffices to prove the result in the case where $D \trianglelefteq G$.

If D is cyclic, all modules in B are D-projective by 2.7.4, and there are only finitely many indecomposable D-modules by 2.2 exercise 1.

If D is non-cyclic, then by 2.5 exercise 1, there are infinitely many isomorphism classes of indecomposable modules for D and so the result follows from 2.12.7 and 2.12.8. □

Exercise

Suppose a Sylow p-subgroup P of G is a t.i. set (i.e. for $g \in G$, either $P^g = P$ or $P \cap P^g = 1$), with normalizer N. Show that Green correspondence gives a one-one correspondence between non-projective indecomposable ΓG-modules and non-projective indecomposable ΓN-modules.

2.13 Semisimplicity of A(G,Triv).

We are still concerned with representation theory over $\Gamma \in \{R,\bar{R}\}$.

In order to show that A(G,Triv) is semisimple, we shall first construct some species for it, and then demonstrate that the elements of A(G,Triv) are separated by the species we have constructed.

A group H is said to be p-hypoelementary if $H/O_p(H)$ is cyclic (recall that $O_p(H)$ denotes the largest normal p-subgroup of a group H). Let $Hyp_p(G)$ be the collection of all p-hypoelementary subgroups of G.

Let V be a trivial source ΓG-module, and suppose $H \in Hyp_p(G)$. Let $V{\downarrow}_H = V_1 \oplus V_2$, where V_1 is a direct sum of indecomposable modules with vertex $O_p(H)$ and V_2 is a direct sum of indecomposable modules with vertex properly contained in $O_p(H)$. Then $O_p(H)$ acts trivially on V_1, and so V_1 is a module for $H/O_p(H)$. Let b be a Brauer species of $H/O_p(H)$, and define $(s_{H,b},V) = (b,V_1)$. Then clearly $s_{H,b}$ is a species of A(G,Triv).

2.13.1 Proposition

Suppose V and W are trivial source ΓG-modules and $(s_{H,b},V) = (s_{H,b},W)$ for all pairs (H,b). Then $V \cong W$.

Proof

Suppose without loss of generality that V and W have no direct summands in common. Let D be a maximal element of the set of vertices of summands of V and W. Suppose $V{\downarrow}_{N_G(D)} = V' \oplus V''$ and $W{\downarrow}_{N_G(D)} = W' \oplus W''$, where V', W' are sums of modules with vertex D, and V", W" are sums of modules whose vertex does not contain D.

Since $(s_{H,b},V') = (s_{H,b},W')$ for each pair (H,b) with $O_p(H) = D$, V' and W' are projective representations of $N_G(D)/D$, and all Brauer species of $N_G(D)/D$ have the same value on each. Thus by 2.11.3, we have $V' \cong W'$. Let V_o' and W_o' be isomorphic indecomposable direct summands of V' and W'. Thus by 2.12.2, the Green correspondents $g(V_o')$ and $g(W_o')$ are isomorphic direct summands of V and W. This contradiction completes the proof of the proposition. □

2.13.2 Corollary

A(G,Triv) is semisimple, and the $s_{H,b}$ are its species. □

Thus by 2.2.1 and the discussion following it, we have idempotents $e_{H,b} \in A(G,Triv)$ with the property that

$$(s_{H,b}, e_{H',b'}) = \begin{cases} 1 & \text{if } (H,b) \text{ is conjugate to } (H',b') \\ 0 & \text{otherwise .} \end{cases}$$

There is a corresponding direct sum decomposition of A(G)

$$A(G) = \bigoplus_{H,b} A(G).e_{H,b} .$$

In this decomposition, H and b run over conjugacy classes of pairs

(H,b) with $H \in Hyp_p(G)$ and b a Brauer species of $H/O_p(H)$ with origin $H/O_p(H)$.

2.13.3 Proposition

(i) $e_D = \sum_{O_p(H) \leq D} e_{H,b}$ is the idempotent generator for $A(G,D)$.

(ii) $e''_D = \sum_{O_p(H)=D} e_{H,b}$ is the idempotent generator for $A''(G,D)$ (see 2.12.6).

Proof

By 2.12.6(i), the idempotent generator for $A(G,D)$ is the sum of the $e_{H,b}$ lying in it, namely the $e_{H,b}$ for which $O_p(H) \leq D$. Similarly, the idempotent generator for $A'(G,D)$ is the sum of the $e_{H,b}$ for which $O_p(H) < D$. □

2.13.4 Proposition

(i) $A(G) = \sum_{H \in Hyp_p(G)} Im(i_{H,G})$

(ii) $\bigcap_{H \in Hyp_p(G)} Ker(r_{G,H}) = 0$

Proof

By 2.13.1, $\bigcap_{H \in Hyp_p(G)} Ker_{A(G,Triv)}(r_{G,H}) = 0$. Thus by Exercise 1 of 2.6, $A(G,Triv) = \sum_{H \in Hyp_p(G)} i_{H,G}(A(H,Triv))$. Thus

$\sum_{H \in Hyp_p(G)} Im(i_{H,G})$ is an ideal of $A(G)$ containing the identity element, proving (i). Then (ii) follows by 2.2.2. □

2.13.5 Proposition

Let $H \leq G$. Then

(i) The idempotent generator of $Im(i_{H,G})$ is $\sum_{\substack{(K,b) \\ K \in Hyp_p(G) \\ K \leq H}} e_{K,b}$.

(ii) The idempotent generator for $Ker(r_{G,H})$ is $\sum_{\substack{(K,b) \\ K \in Hyp_p(G) \\ K \text{ not conjugate} \\ \text{to a subgroup of } H}} e_{K,b}$.

(the sums run over G-conjugacy classes of pairs (K,b).)

Proof

This follows immediately from 2.13.4 and exercise 1 of 2.6. □

Finally, we prove Conlon's induction theorem.

2.13.6 Theorem (Conlon)

There exist rational numbers $\lambda_H \in \mathbb{Q}$ for each conjugacy class of p-hypoelementary subgroup H such that

$$1_G = \sum_{H \in \mathrm{Hyp}_p(G)} \lambda_H 1_H\uparrow^G$$

(the sum runs over a set of representatives of conjugacy classes of p-hypoelementary subgroups).

Proof

Let Θ_G be the $\mathbb{Q}$-linear span in $A(G,\mathrm{Triv})$ of the permutation modules $1_H\uparrow^G$. Then exactly as in 2.2.2, for any subgroup H we have

$$\Theta_G = i_{H,G}(\Theta_H) \oplus \mathrm{Ker}_{\Theta_G}(r_{G,H}) .$$

Now by 2.13.4,

$$\bigcap_{H \in \mathrm{Hyp}_p(G)} \mathrm{Ker}_{\Theta_G}(r_{G,H}) = 0$$

and so

$$\sum_{H \in \mathrm{Hyp}_p(G)} i_{H,G}(\Theta_H) = \Theta_G$$

as required. □

Exercise

Use 2.13.3 and 2.13.5 to show that if $N \trianglelefteq G$ has index p^n, then $A(G,N) = \mathrm{Im}(i_{N,G})$. Deduce that $a(G,N)/i_{N,G}(a(N))$ is a (possibly infinite) p-group of exponent dividing p^n.

In fact, Green has shown [54] that if k is algebraically closed then $a(G,N) = i_{N,G}(a(N))$.

2.14 Structure Theorem for Vertices and Origins

This section consists of just one theorem describing the nature of the vertices and origins of a species. The proof of the theorem is a good illustration of the influence which the subring $A(G,\mathrm{Triv})$ exerts on the structure of $A(G)$.

2.14.1 Theorem

Let s be a species of $A_\Gamma(G)$, $\Gamma \in \{R,\bar{R}\}$. Then

(i) All origins of s are conjugate.

(ii) All the vertices of s are conjugate.

Let H be an origin of s. Then

(iii) H is p-hypoelementary.

(iv) $O_p(H)$ is a vertex of s.

Proof

(i) is proved in 2.9.3.

(ii) is proved in 2.5.2(ii).

(iii) Suppose H is an origin of s. By 2.13.4, we have

$$\mathrm{Im}(i_{H,G}) = \sum_{\substack{K \in \mathrm{Hyp}_p(G) \\ K \leq H}} \mathrm{Im}(i_{K,G}) .$$

By 2.9.2, $\mathrm{Ker}(s) \not\geq \mathrm{Im}(i_{H,G})$, and so for some $K \in \mathrm{Hyp}_p(G)$ with $K \leq H$, $\mathrm{Ker}(s) \not\geq \mathrm{Im}(i_{K,G})$. By minimality of H, $K = H$.

(iv) Since H is an origin of s, by 2.13.5 s does not vanish on $\sum_{\substack{(K,b) \\ K \in \mathrm{Hyp}_p(G) \\ K \leq H}} e_{K,b}$, but does vanish on $\sum_{\substack{(K,b) \\ K \in \mathrm{Hyp}_p(G) \\ K < H}} e_{K,b}$. Thus s does not vanish on some $e_{H,b}$, and so by 2.13.3, s does not vanish on $e''_{O_p(H)}$. Thus s does not vanish on $e_{O_p(H)}$, but does vanish on $e_{D'}$ for every $D' < O_p(H)$. Hence $O_p(H)$ is a vertex of s. □

2.15 Tensor Induction, and Yet Another Decomposition of A(G)

In this section, we introduce the notion of tensor induction, and use it to prove that for any permutation representation S of a group G, we have the decomposition $A(G) = A(G,S) \oplus A_0(G,S)$ (see Theorem 2.15.6), as promised in section 2.3. The proof of this theorem is another good illustration of the influence which $A(G,\mathrm{Triv})$ exerts on the structure of $A(G)$.

Suppose $H \leq G$ and V is a ΓH-module. Then $V \otimes_{\Gamma H} \Gamma G$ splits naturally (as a vector space) as a direct sum of blocks $V \otimes g_i$, for g_i a set of right coset representatives of H in G, and G permutes these blocks in the same way as it permutes the right cosets of H. Thus $\bigotimes_i (V \otimes g_i)$ has a natural structure as a ΓG-module, and is written $V \mathbin{\hat{\otimes}}^G$, and called ' V tensor induced up to G '. The basic properties of tensor induction are as follows.

2.15.1 Lemma

Let $H \leq G$, and let V_1 and V_2 be ΓH-modules.

(i) $(V_1 \otimes V_2) \mathbin{\hat{\otimes}}^G \cong V_1 \mathbin{\hat{\otimes}}^G \otimes V_2 \mathbin{\hat{\otimes}}^G$.

(ii) $(V_1 \oplus V_2) \overset{\uparrow}{\otimes}{}^{G} \cong V_1 \overset{\uparrow}{\otimes}{}^{G} \oplus V_2 \overset{\uparrow}{\otimes}{}^{G} \oplus X$ with $X \in \sum_{K<G} \mathrm{Im}(i_{K,G})$.

(iii) If $H' \leq H$ and W is a $\Gamma H'$-module, then

$$W{\uparrow}^{H} \overset{\uparrow}{\otimes}{}^{G} \in \sum_{\substack{K \leq G \\ K \cap H \leq H'}} \mathrm{Im}(i_{K,G}) .$$

(iv) $\overset{\uparrow}{\otimes}$ induces a ring homomorphism

$$i^{\otimes}_{H,G}\colon A(H)/\sum_{K<H} \mathrm{Im}(i_{K,H}) \to A(G)/\sum_{K<G} \mathrm{Im}(i_{K,G})$$

which takes the identity element to the identity element.

(v) If $0 \to V' \to V \to V'' \to 0$ is an S-split short exact sequence of ΓH-modules, then

$$V \overset{\uparrow}{\otimes}{}^{G} \otimes \Gamma S \cong (V' \oplus V'') \overset{\uparrow}{\otimes}{}^{G} \otimes \Gamma S.$$

<u>Proof</u>

(i) is clear from the definition.

(ii) $$(V_1 \oplus V_2) \overset{\uparrow}{\otimes}{}^{G} = \bigotimes_i ((V_1 \oplus V_2) \otimes g_i)$$

$$= \bigotimes_i (V_1 \otimes g_i) \oplus \bigotimes_i (V_2 \otimes g_i) \oplus X$$

where $$X = \bigoplus_{\substack{j_i=1,2 \\ \text{not all } j_i \\ \text{equal}}} \left(\bigotimes_i (V_{j_i} \otimes g_i)\right) .$$

Thus, as a ΓG-module, X splits as a direct sum of submodules corresponding to the G-orbits of ways of choosing the j_i's. Each such summand is a module induced from the stabilizer of such a choice.

(iii) Let g_i be coset representatives of H in G, and h_j be coset representatives of H' in H. Then

$$W{\uparrow}^{H} \overset{\uparrow}{\otimes}{}^{G} = \bigotimes_i \left(\left(\bigoplus_j (W \otimes h_j)\right) \otimes g_i\right)$$

$$= \bigoplus_{\substack{\text{possible} \\ \text{choices of} \\ \text{one } h_{j_i} \text{ for} \\ \text{each } i}} \left(\bigotimes_i (W \otimes h_{j_i} g_i)\right).$$

Thus as a ΓG-module, $W{\uparrow}^{H} \overset{\uparrow}{\otimes}{}^{G}$ splits as a direct sum of submodules corresponding to the G-orbits of ways of choosing the j_i's. Each such summand is a module induced from the stabilizer of such a choice, and the elements of H stabilizing such a choice are contained in an H-conjugate of H'.

(iv) This follows immediately from (i), (ii) and (iii).

(v) As a ΓG-module, $V \hat{\otimes}^{G}$ has a natural filtration $V' \hat{\otimes}^{G} = U_0 \le U_1 \le \dots \le U_n = V \hat{\otimes}^{G}$ $(n = |G:H|)$ where

$$U_j = \langle (\bigotimes_{i \notin J} v_i' \otimes g_i) \otimes (\bigotimes_{i \in J} v_i \otimes g_i), \text{ for } J \text{ a subset of size } j \text{ of the right cosets of } H \text{ in } G, \text{ and } v_i' \in V', v_i \in V\rangle .$$

It is easily seen that

$$(V' \oplus V'') \hat{\otimes}^{G} \cong U_0 \oplus \bigoplus_{j=1}^{n} U_j/U_{j-1} .$$

Thus we must show that tensoring with ΓS splits the filtration $V' \hat{\otimes}^{G} = U_0 \le U_1 \le \dots \le U_n = V \hat{\otimes}^{G}$; i.e. we must find a left inverse Ψ_j for the natural map

$$\Phi_j : U_j \otimes \Gamma S \to (U_j \otimes \Gamma S)/(U_{j-1} \otimes \Gamma S) \cong (U_j/U_{j-1}) \otimes \Gamma S.$$

Suppose $f: V'' \otimes \Gamma S \to V \otimes \Gamma S$ is an S-splitting for $0 \to V' \to V \to V'' \to 0$, and let

$$(\sum_{x \in S} v_x'' \otimes x)f = \sum_{x \in S} (v_x'' f_x) \otimes x$$

with the f_x linear maps from V'' to V. Since f is a ΓH-module homomorphism, we get $f_x h = h f_{xh}$ for $h \in H$.

The typical generator for $(U_j \otimes \Gamma S)/(U_{j-1} \otimes \Gamma S)$ is $((\bigotimes_{i \notin J} (v_i' \otimes g_i)) \otimes (\bigotimes_{i \in J} (v_i'' \otimes g_i))) \otimes x$. We define Ψ_j to be the map sending this generator to $((\bigotimes_{i \notin J} (v_i' \otimes g_i)) \otimes (\bigotimes_{i \in J} (v_i'' f_{xg_i^{-1}} \otimes g_i))) \otimes x$.

It is easily checked that Ψ_j is a ΓG-module homomorphism left inverse to Φ_j. □

<u>Remark</u>

When trying to prove, for a group G, that $A(G) = A(G,S) \oplus A_0(G,S)$, we only have to show that $1 = \alpha + \beta \in A(G,S) + A_0(G,S)$. This is because $A(G,S)$ and $A_0(G,S)$ are ideals of $A(G)$ whose product is zero, and so if $x \in A(G,S) \cap A_0(G,S)$ then

$$x = x.1 = x.\alpha + x.\beta = 0 .$$

2.15.2 <u>Lemma</u>

Suppose D is a p-group and S is a permutation representation of D with $D \notin \mathrm{Fix}_D(S)$. Then

$$A(D) = A_o(D,S) + \sum_{K<D} \mathrm{Im}(i_{K,D}).$$

Proof

Without loss of generality D acts transitively on S. Let D' be the stabilizer of a point in S, and D'' a maximal subgroup of D containing D'. Then $A_o(D,S) \geq A_o(D,D'')$. Since the group algebra of D/D'' is indecomposable (see 2.1 exercise 1), $A(D/D'',1) = \mathrm{Im}(i_{1,D/D''})$ is one-dimensional. Thus

$$\begin{aligned} 1_{A(D)} \in A(D/D'') &= A_o(D/D'',1) \oplus A(D/D'',1) \qquad \text{by 2.11.3} \\ &= A_o(D/D'',1) \oplus \mathrm{Im}(i_{1,D/D''}) \\ &\subseteq A_o(D/D'') + \mathrm{Im}(i_{D'',D}) \\ &\subseteq A_o(D,S) + \sum_{K<H} \mathrm{Im}(i_{K,D}). \quad \square \end{aligned}$$

2.15.3 Lemma

Suppose $H \leq G$ and $O_p(H) \notin \mathrm{Fix}_G(S)$. Then

$$A(H) = A_o(H,S) + \sum_{K<H} \mathrm{Im}(i_{K,H}).$$

Proof

Let $D = O_p(H) \notin \mathrm{Fix}_G(S)$. Then by 2.15.1 and 2.15.2

$$i^{\otimes}_{D,H}: A(D)/\sum_{K<D} \mathrm{Im}(i_{K,D}) = \frac{A_o(D,S) + \sum_{K<D} \mathrm{Im}(i_{K,D})}{\sum_{K<D} \mathrm{Im}(i_{K,D})} \to \frac{A(H)}{\sum_{K<H} \mathrm{Im}(i_{K,H})}$$

takes the identity element to the identity element. But it also takes $A_o(D,S)$ into $A_o(H,S) + \sum_{K<H} \mathrm{Im}(i_{K,H})$ by 2.15.1 (v) and (ii), and so

$$1_{A(H)} \in A_o(H,S) + \sum_{K<H} \mathrm{Im}(i_{K,H}). \quad \square$$

2.15.4 Lemma

If $H \in \mathrm{Hyp}_p(G)$ then $A(H) = A(H,S) \oplus A_o(H,S)$.

Proof

If $O_p(H) \in \mathrm{Fix}_H(S)$ then $A(H) = A(H,S)$, by 2.3.3. If $O_p(H) \notin \mathrm{Fix}_H(S)$, then by 2.15.3, $A(H) = A_o(H,S) + \sum_{K<H} \mathrm{Im}(i_{K,H})$. By induction, for each $K < H$, $A(K) = A(K,S) \oplus A_o(K,S)$, and so by 2.3.8, $\mathrm{Im}(i_{K,H}) \subseteq A(H,S) + A_o(H,S)$. $\square$

2.15.5 Theorem

For any group G and permutation representation S, we have

$A(G) = A(G,S) \oplus A_0(G,S)$.

Proof

By 2.13.4(i), $A(G) = \sum_{H \varepsilon \mathrm{Hyp}_p(G)} \mathrm{Im}(i_{H,G})$. By 2.15.4 and 2.3.8, $\mathrm{Im}(i_{H,G}) \subseteq A(G,S) + A_0(G,S)$ for $H \varepsilon \mathrm{Hyp}_p(G)$. □

2.15.6 Corollary

If $H \leq G$ then $A(G) = A(G,H) \oplus A_0(G,H)$. □

2.15.7 Corollary

If s is a species of $A(G)$, then D contains a vertex of s if and only if for every D-split short exact sequence $0 \to V' \to V \to V'' \to 0$ we have

$$(s,V) = (s,V') + (s,V''). \quad \square$$

Remark

Dress [45] has shown that in fact $\hat{a}(G) = \hat{a}(G,H) \oplus \hat{a}_0(G,H)$; i.e. $A(G)/(a(G,H) + a_0(G,H))$ is a p-torsion group, cf. 2.16.5.

2.16 Power Maps on A(G)

In this section we construct maps $\psi^n: a_k(G) \to a_k(G)$ called the power maps. These are ring homomorphisms, and have the property that if b_g is a Brauer species then $(b_{g^n},x) = (b_g, \psi^n(x))$. These are the modular analogues of what are called the Adams operations in ordinary representation theory. We shall use these maps to construct the powers of a general species, and we shall investigate the origins and vertices of the powers of a species.

We begin by constructing the operators ψ^n in the case where n is coprime to p. Let n be a natural number coprime to p, and let $T = \langle \alpha: \alpha^n = 1 \rangle$ be a cyclic group of order n. Let ε be a primitive $n^{\underline{th}}$ root of unity in the algebraic closure of k and let η be a primitive $n^{\underline{th}}$ root of unity in $\mathbb{C}$. If X is a module for $T \times G$, then we denote by X_{ε^i} the eigenspace of α on X with eigenvalue ε^i. Then X_{ε^i} is a $T \times G$ -invariant direct summand of X, and $X = \bigoplus_{i=1}^{n} X_{\varepsilon^i}$.

Now let V be an kG-module. Then $[V \hat{\otimes}^{T \times G}]_{\varepsilon^i}$ restricts to a kG-module, and whenever $\langle \alpha^i \rangle = \langle \alpha^j \rangle$, we have $[V \hat{\otimes}^{T \times G}]_{\varepsilon^i} \cong [V \hat{\otimes}^{T\times G}]_{\varepsilon^j}$ as kG-modules. We define

$$\psi^n(V) = \sum_{i=1}^{n} \eta^i [V \otimes^{T \times G}]_{\varepsilon^i} \in A(G).$$

2.16.1 Proposition

If V_1 and V_2 are kG-modules then

(i) $\psi^n(V_1 \oplus V_2) = \psi^n(V_1) + \psi^n(V_2)$

(ii) $\psi^n(V_1 \otimes V_2) = \psi^n(V_1)\psi^n(V_2)$.

Proof

(i) As a module for G, we have

$$\otimes^n(V_1 \oplus V_2) = \bigoplus_{\substack{i_1=1,2 \\ \vdots \\ i_n=1,2}} (V_{i_1} \otimes \dots \otimes V_{i_n}).$$

Under the action of T, there are two fixed summands, $\otimes^n(V_1)$ and $\otimes^n(V_2)$. Apart from these, each orbit forms a module for $T \times G$ of the form $Y \otimes Z$, where Y is a permutation module for T on the cosets of a proper subgroup. Thus as an element of $A(G)$,

$$\sum_{i=1}^{n} \eta^i [Y \otimes Z]_{\varepsilon^i} = 0.$$

Hence the result.

(ii) $\otimes^n(V_1 \otimes V_2) = \otimes^n(V_1) \otimes \otimes^n(V_2)$.

Hence

$$[\otimes^n(V_1 \otimes V_2)]_{\varepsilon^i} = \sum_{j=1}^{n} [\otimes^n(V_1)]_{\varepsilon^j} [\otimes^n(V_2)]_{\varepsilon^{i-j}} .$$

Thus we have

$$\begin{aligned} \psi^n(V_1 \otimes V_2) &= \sum_{i=1}^{n} \eta^i [\otimes^n(V_1 \otimes V_2)]_{\varepsilon^i} \\ &= \sum_{i,j=1}^{n} \eta^j [\otimes^n(V_1)]_{\varepsilon^j} \cdot \eta^{i-j} [\otimes^n(V_2)]_{\varepsilon^{i-j}} \\ &= \psi^n(V_1)\,\psi^n(V_2) . \quad \square \end{aligned}$$

By 2.16.1, we may extend ψ^n linearly to give a ring endomorphism of $A(G)$. In fact, the image under ψ^n of an element of $a(G)$ is in $a(G)$, as the following proposition shows.

2.16.2 Proposition

For d dividing n, let ε_d be a primitive d^{th} root of unity in the algebraic closure of k. Then

$$\psi^n(V) = \sum_{d|n} \mu(d) [V \otimes^{T \times G}]_{\varepsilon_d} .$$

(Here, μ is the Möbius function of multiplicative number theory)

Proof

Whenever $<\alpha^i> = <\alpha^j>$, $[V \overset{\uparrow T\times G}{\otimes}]_{\varepsilon^i} \cong [V \overset{\uparrow T\times G}{\otimes}]_{\varepsilon^j}$. Thus

$$\psi^n(V) = \sum_{i=1}^{n} \eta^i [V \overset{\uparrow T\times G}{\otimes}]_{\varepsilon^i}$$

$$= \sum_{d|n} \Big(\sum_{\substack{(i,d)=1 \\ 1\le i\le d}} \varepsilon_d^i \Big) [V \overset{\uparrow T\times G}{\otimes}]_{\varepsilon_d}$$

$$= \sum_{d|n} \mu(d) [V \overset{\uparrow T\times G}{\otimes}]_{\varepsilon_d} . \qquad \square$$

Example

If $p \neq 2$, we have

$$\psi^2(V) = S^2(V) - \Lambda^2(V).$$

Thus, in particular, if V is irreducible then the Frobenius-Schur indicator is defined by

$$\mathrm{Ind}(V) = (1, \psi^2(V)) = \begin{cases} +1 & \text{if } V \text{ is orthogonal} \\ -1 & \text{if } V \text{ is symplectic} \\ 0 & \text{otherwise.} \end{cases}$$

(Recall that $(\, , \,)$ is the inner product on $A(G)$ given by bilinearly extending $(U,V) = \dim_k \mathrm{Hom}_{kG}(U,V)$).

2.16.3 Definition

We define the nth power of a species s of $A(G)$, for $n \in \mathbb{N}\backslash p\mathbb{N}$, via

$$(s^n, x) = (s, \psi^n(x)).$$

Proposition 2.16.1 shows that s^n is again a species of $A(G)$.

2.16.4 Proposition

If b is a Brauer species of $A(G)$ (see section 2.11) corresponding to a p'-element g, then b^n is the Brauer species corresponding to g^n.

Proof

Let V be a kG-module, and let b' be the Brauer species corresponding to g^n. We may choose a basis $v_1, \dots, v_r$ of V consisting of eigenvectors of g. Let $v_i g = \lambda_i v_i$. Then as $k<g>$-modules, $V = \oplus < v_i >$, and so

$$(b^n, V) = (b, \psi^n(V)) = (b, \sum_{i=1}^{r} \psi^n(< v_i >)) = \sum_{i=1}^{r} (b, \psi^n(< v_i >))$$

$$= \sum_{i=1}^{r} \lambda_i^n = (b', V). \quad \square$$

As our first application, we give Kervaire's proof of a theorem of Brauer.

2.16.5 Theorem

The determinant of the Cartan matrix is a power of p.

Proof

This is the same as saying that the cokernel of the Cartan homomorphism

$$c: a(G,1) \to a(G) \to a(G)/a_o(G,1)$$

is a p-group.

Let m be the p'-part of the exponent of G. If $x \in a(G,1)$ then $\psi^m(x)$ is an element of $a(G,1)$ by 2.16.2, and any Brauer species has value $\dim(x)$ on $\psi^m(x)$ by 2.16.4. Thus by 2.11.3, if $x \in a(G,1)$ then $\dim(x).1 \in \mathrm{Im}(c)$.

For each prime $q \neq p$ dividing $|G|$, let Q be a Sylow q-subgroup of G. Then $1_Q\uparrow^G \in a(G,1)$ since it is induced from a projective kQ-module. Thus $|G:Q|.1 \in \mathrm{Im}(c)$. It now follows from the Chinese remainder theorem that $|G|_p.1 \in \mathrm{Im}(c)$, where $|G|_p$ is the p-part of the order of G. Hence if $x \in a(G)/a_o(G,1)$ then $|G|_p.x \in \mathrm{Im}(c)$, and the theorem is proved. □

Remark

This could be rephrased as saying that $a(G)/(a(G,1) + a_o(G,1))$ is a p-torsion group; see the remark after 2.15.7.

We now wish to prove that $\psi^m\psi^n = \psi^{mn}$. We start off with a lemma.

2.16.6 Lemma

Let S_n denote the symmetric group on n letters. Then there is a subgroup T_n of S_n having the following properties.

(i) T_n contains a cyclic group of order n which is transitive on the n letters.

(ii) If $n = n_1n_2$ then T_n contains the direct product of the cyclic groups of orders n_1 and n_2, in its direct product action on the n letters.

(iii) If a prime q divides $|T_n|$ then q also divides n.

Proof

Let $n = \Pi p_i^{\alpha_i}$. Then we have a subgroup

$$\Pi S_{p_i^{\alpha_i}} \leq S_n$$

with direct product action on the n points. Let P_i be a Sylow

p_i-subgroup of $S_{p_i^{\alpha_i}}$, and let

$$T_n = \Pi P_i \le \Pi S_{p_i^{\alpha_i}} .$$

Then properties (i) and (iii) are clearly satisfied. To check property (ii), let $n = n_1 n_2$ with $n_1 = \Pi p_i^{\beta_i}$, $n_2 = \Pi p_i^{\gamma_i}$, and $\beta_i + \gamma_i = \alpha_i$. Let $Q_i \times R_i$ denote a Sylow p-subgroup of $S_{p_i^{\beta_i}} \times S_{p_i^{\gamma_i}} \le S_{p_i^{\alpha_i}}$, with $Q_i \times R_i \le P_i$. Then $\Pi Q_i \times \Pi R_i \le S_{n_1} \times S_{n_2}$ contains the appropriate direct product of cyclic groups. □

2.16.7 Theorem

$$\psi^m \psi^n = \psi^{mn}$$

Proof

Without loss of generality, we may assume that k is a splitting field for T_{mn} (see 2.2 exercise 4). Thus by property (iii), p does not divide $|T_{mn}|$, and so the representation theory of kT_{mn} is the same as the representation theory of $\mathbb{C}T_{mn}$. In particular, the central idempotents of kT_{mn} are in natural one-one correspondence with the central idempotents of $\mathbb{C}T_{mn}$, and kT_{mn} is semisimple.

By properties (i) and (ii) of T_{mn}, and the definition of the ψ operators, $\psi^m\psi^n(V)$ and $\psi^{mn}(V)$ are of the form $\Sigma\, \lambda_i(\otimes^{mn}(V).e_i)$ and $\Sigma\, \lambda_i'(\otimes^{mn}(V).e_i)$, where the e_i are the primitive central idempotents of T_{mn}, and the λ_i and λ_i' are independent of V. Moreover, the λ_i and λ_i' may both be expressed in terms of induced characters from the subgroups of T_{mn} given in the definition, and hence if we keep m and n constant and vary p over primes not dividing mn, the λ_i and λ_i' do not vary. Thus it is sufficient to prove the result in the case where p divides neither mn nor $|G|$. In this case, every species is a Brauer species, and modules are characterized by the values of Brauer species. By 2.16.4, we have

$$\begin{aligned}(b,\ \psi^m\psi^n(V)) &= (b^m, \psi^n(V)) \\ &= ((b^m)^n, V) \\ &= (b^{mn}, V) \\ &= (b,\ \psi^{mn}(V)) .\end{aligned}$$

Thus the λ_i and λ_i' are equal, and the result is proved. □

We now extend the definition of ψ^n to include all $n \in \mathbb{N}$ as follows. Let F denote the Frobenius map on $a(G)$ or $A(G)$. Thus if V is a module, $F(V)$ is the module with the same addition and same group action, but with scalar multiplication defined by first raising the field element to the pth power, and then applying the old scalar multiplication. The map F commutes with ψ^n for n coprime to p, and so we may define, for any $n \in \mathbb{N}$ with $n = p^a.n_o$ and n_o coprime to p,

$$\psi^n(V) = F^a\, \psi^{n_o}(V)\ .$$

It is easy to check that propositions 2.16.1 and 2.16.4, and theorem 2.16.7 remain valid with the definition, and so we extend definition 2.16.3 appropriately.

Remark

If we define

$$\lambda^n(x) = \frac{1}{n!}\begin{vmatrix} \psi^1(x) & 1 & & & \\ \psi^2(x) & \psi^1(x) & 2 & & \text{\huge 0} \\ \psi^3(x) & \psi^2(x) & \psi^1(x) & 3 & \\ \vdots & & & \ddots & \\ & & & & n-1 \\ \psi^n(x) & \cdots & & & \psi^1(x) \end{vmatrix}$$

then these λ-operations make $A(G)$ into a special λ-ring (see [62]). In fact the subring $\hat{a}(G) = a(G) \underset{\mathbb{Z}}{\otimes} \mathbb{Z}(\frac{1}{p})$ is stable under these operations, see [14].

Next, we examine the effects of ψ^n on origins and vertices of species.

2.16.8 Definition

If H is a p-hypoelementary group and $n = p^a.n_o$ with n_o coprime to p, we let $H^{[n]}$ denote the unique subgroup of index $(|H|,n_o)$ in H.

2.16.9 Lemma

Let $s_{H,b}$ be as in section 2.13. Then

$$(s_{H,b})^n = s_{H^{[n]},b^n}\ .$$

Proof

Let V be a trivial source $\bar{R}G$-module and let $V{\downarrow}_H = W_1 \oplus W_2$, where W_1 is a direct sum of modules with vertex $O_p(H)$ and $W_2 \in A'(G,H)$. Then by 2.16.1, $\psi^n(V){\downarrow}_H = \psi^n(V{\downarrow}_H) = \psi^n(W_1) + \psi^n(W_2)$, $\psi^n(W_1)$ is a linear combination of trivial source modules with vertex $O_p(H)$, and $\psi^n(W_2) \in A'(G,H)$. Thus

$$\begin{aligned}((s_{H,b})^n, V) &= (s_{H,b}, \psi^n(V)) \\ &= (b, \psi^n(W_1)) \\ &= (b^n, W_1) \\ &= (s_{H^{[n]},b^n}, V). \qquad \square\end{aligned}$$

2.16.10 Lemma

$$\psi^n(e_{H,b}) = \Sigma\, e_{H',b'}$$

where the sum runs over one representative of each G-conjugacy class of pairs (H',b') with $(H')^{[n]} = H$ and $(b')^n = b$.

Proof

$$\psi^n(e_{H,b}) = \sum_{\substack{\text{all} \\ (H',b')}} (s_{H',b'}, \psi^n(e_{H,b})).e_{H',b'} \quad .$$

(Here, the sum runs over one representative of each G-conjugacy class of pairs (H',b').)

$$= \Sigma\ (s_{(H')^{[n]},(b')^n}, e_{H,b}).e_{H',b'}$$

by lemma 2.16.9.

Thus the coefficient of $e_{H',b'}$ is one if $((H')^{[n]},(b')^n)$ is G-conjugate to (H,b) and zero otherwise. $\square$

2.16.11 Theorem

(i) If H is an origin of s, then $H^{[n]}$ is an origin of s^n.

(ii) If D is a vertex of s, then D is also a vertex of s^n.

Proof

(i) If H is an origin of s, then for some Brauer species b of $H/O_p(H)$ with origin $H/O_p(H)$, $(s,e_{H,b}) = 1$. Thus by lemma 2.16.10,

$$(s^n, e_{H^{[n]},b^n}) = (s, \psi^n(e_{H^{[n]},b^n})) = 1.$$

Hence $H^{[n]}$ is an origin for s^n.

(ii) By 2.14.1, we may take $D = O_p(H)$, and the result follows from (i). □

2.17 Almost Split Sequences

In this section, we construct certain short exact sequences, called almost split sequences, of modules for the group algebra kG. These were first constructed by Auslander and Reiten [5] in the more general context of modules for an Artin algebra. We shall restrict our attention to group algebras, since this makes the arguments easier to follow. It turns out that the existence of these sequences depends upon an interesting identity, namely Theorem 2.17.5. The reader happy with abstract categorical arguments should also see Gabriel's impressively short proof of the existence of almost split sequences for arbitrary Artin algebras in [53]. See also the remark at the end of 2.17.

In the next section we shall see that these short exact sequences play an important role in the structure of A(G), namely they give us certain 'dual elements' under the inner products investigated in section 2.4, to the basis of A(G) consisting of indecomposable modules.

2.17.1 Lemma

Let U and V be kG-modules. Then there is a natural duality $((U,V)^{1,G})^* \cong (V, \Omega U)^{1,G}$

Proof

By 2.1.1(iv), we have

$$(U,V)^{1,G} \cong (V^* \otimes U,k)^{1,G}$$

and

$$(V, \Omega U)^{1,G} \cong (k,V^* \otimes \Omega U)^{1,G} .$$

By Schanuel's lemma, $V^* \otimes \Omega U \cong \Omega(V^* \otimes U) \oplus P$ for some projective module P, and so by 2.3.4,

$$(k,V^* \otimes \Omega U)^{1,G} \cong (k, \Omega(V^* \otimes U))^{1,G} .$$

Thus we must show that

$$(k, \Omega(V^* \otimes U))^{1,G} \cong ((V^* \otimes U,k)^{1,G})^* .$$

In fact we shall show that for any module X,

$$(k, \Omega X)^{1,G} \cong ((X,k)^{1,G})^* .$$

Without loss of generality, X has no projective direct summands. Let P_X be the projective cover of X. Then since kG is a symmetric algebra, there are as many copies of the trivial module in the head as in the socle of P_X, and there is a natural isomorphism between the spaces given by multiplication by $\sum_{g \in G} g$.

Thus $(k, \Omega X)^{1,G} \cong (k,P_X)^G \cong ((P_X,k)^G)^* \cong ((X,k)^{1,G})^*$. □

Applying 2.17.1 twice, we get

2.17.2 Corollary (Feit)

$(U,V)^{1,G} \cong (\Omega U,\Omega V)^{1,G}$. □

The isomorphism of 2.17.2 may be given the following interpretation. Let P_U and P_V be the projective covers of U and V. Then any map from U to V lifts to a map from P_U to P_V, and the image of ΩU lies in ΩV. Thus we obtain a (not necessarily unique) map from ΩU to ΩV. However, if a map from U to V factors through a projective module, then so does the induced map from ΩU to ΩV, and vice-versa. Thus we get a well defined injection, and since the spaces are of equal dimension, this is an isomorphism.

In particular, when U = V, we write $\underline{\mathrm{End}}_{kG}(U)$ for $(U,U)^{1,G}$. The above map from $\underline{\mathrm{End}}_{kG}(U)$ to $\underline{\mathrm{End}}_{kG}(\Omega U)$ clearly preserves composition of endomorphisms, and so we have the following result.

2.17.3 Proposition

There is a natural ring isomorphism

$$\underline{\mathrm{End}}_{kG}(U) \cong \underline{\mathrm{End}}_{kG}(\Omega U). \quad □$$

But this means that both sides of 2.17.1 are $\underline{\mathrm{End}}_{kG}(V)$-$\underline{\mathrm{End}}_{kG}(U)$ bimodules. Is the given isomorphism a bimodule isomorphism? Clearly the first two isomorphisms given in the proof are bimodule isomorphisms. So we simply need the following proposition.

2.17.4 Proposition

There is an $\underline{\mathrm{End}}_{kG}(V)$-$\underline{\mathrm{End}}_{kG}(U)$ bimodule isomorphism

$$(V^* \otimes U,k)^{1,G} \cong ((V^* \otimes \Omega U)^{1,G})^* ,$$

namely the map induced by the map

$$\gamma\colon (V^* \otimes U,k)^G \to ((V^* \otimes \Omega U)^{1,G})^*$$

given as follows. For $y^* \in V^*$, $x \in P_U$ and $\varphi \in (V^* \otimes U,k)^G$,

$$((y^* \otimes x)(\sum_{g \in G} g) + (V^* \otimes \Omega U)_1^G)(\varphi\gamma) = (y^* \otimes x + V^* \otimes \Omega U)\varphi .$$

Proof

$(V^* \otimes \Omega U)^G$ is spanned by elements of the form $(y^* \otimes x)(\sum_{g \in G} g)$ with $y^* \otimes x \in V^* \otimes P_U$. To see that γ is well defined, let a_1, $a_2 \in V^* \otimes P_U$ with $(a_1 - a_2)(\sum_{g \in G} g) \in (V^* \otimes \Omega U)_1^G$. Then any homomorphism from $V^* \otimes P_U$ to k with $V^* \otimes \Omega U$ in its kernel has $a_1 - a_2$ in its kernel. γ is clearly surjective, and $(V^* \otimes U,k)_1^G \subseteq \mathrm{Ker}(\gamma)$. By 2.17.1, $\dim(V^* \otimes \Omega U)^{1,G})^* = \dim(V^* \otimes U,k)^{1,G}$, and so γ induces an isomorphism, which clearly preserves the **bimodule** structure. □

2.17.5 Theorem

There are $\underline{\mathrm{End}}_{kG}(V)$-$\underline{\mathrm{End}}_{kG}(U)$ bimodule isomorphisms

$$((U,V)^{1,G})^* \cong (V, \Omega U)^{1,G} \cong \mathrm{Ext}_G^1(V, \Omega^2 U)$$

Proof

We have already proved the first isomorphism. To prove the second, we have a short exact sequence

$$0 \to \Omega^2 U \to P_{\Omega U} \to \Omega U \to 0$$

which gives rise to a long exact sequence (see 1.4)

$$0 \to (V, \Omega^2 U)^G \to (V, P_{\Omega U})^G \to (V, \Omega U)^G \to \mathrm{Ext}_G^1(V, \Omega^2 U)^G \to \mathrm{Ext}_G^1(V, P_{\Omega U})^G = 0.$$

Since the image of $(V,P_{\Omega U})^G$ in $(V,\Omega U)^G$ is exactly $(V,\Omega U)_1^G$, the second isomorphism follows. The naturality of the long exact sequence means that this is a bimodule isomorphism. □

We are now ready to examine the almost split sequences.

2.17.6 Definition

An almost split sequence or Auslander-Reiten sequence is a short exact sequence of modules $0 \to A \to B \xrightarrow{\sigma} C \to 0$ satisfying the following conditions.

(i) A and C are indecomposable.

(ii) σ does not split.

(iii) If $\rho: D \to C$ is not a split epimorphism (i.e. unless C is isomorphic to a direct summand of D and ρ is the projection) then ρ factors through σ.

Auslander and Reiten proved for a general Artin algebra that for each non-projective C there is a unique almost split sequence terminating in C, and gave a recipe for obtaining the module A. It turns out that for group algebras $A \cong \Omega^2 C$, as is seen in the following theorem.

2.17.7 Theorem

Let C be a non-projective indecomposable kG-module. Then there exists an almost split sequence terminating in C. This sequence is unique up to isomorphism of short exact sequences, and its first term is isomorphic to $\Omega^2 C$.

Proof

(i) We first prove existence. By 2.17.5, we have

$$((C,C)^{1,G})^* \cong \mathrm{Ext}^1_G(C, \Omega^2 C).$$

If C is not projective, then $(C,C)^{1,G}$ is a local ring by 1.3.3, since $(C,C)^G_1$ is not the whole of $(C,C)^G$. Thus as an $\underline{\mathrm{End}}_{kG}(C)$-$\underline{\mathrm{End}}_{kG}(C)$ bimodule, or even as a one sided module, $\mathrm{Ext}^1_G(C,\Omega^2 C)$ has an irreducible socle, and any two extensions generating this socle are equivalent under an automorphism of C, and hence give rise to equivalent short exact sequences. We claim that a generator $0 \to \Omega^2 C \to X_C \overset{\sigma}{\to} C \to 0$ for $\mathrm{Soc}\,\mathrm{Ext}^1_G(C, \Omega^2 C)$ has the desired properties. Clearly properties (i) and (ii) are satisfied, so we must check (iii).

Let $\gamma : \Omega C \to \Omega^2 C$ with image in $(\Omega C, \Omega^2 C)^{1,G} \cong \mathrm{Ext}^1_G(C, \Omega^2 C)$ our generator for $\mathrm{Soc}\,\mathrm{Ext}^1_G(C,\Omega^2 C)$. Thus our short exact sequence is a pushout of the form

$$\begin{array}{ccccccccc}
0 & \longrightarrow & \Omega C & \longrightarrow & P_C & \longrightarrow & C & \longrightarrow & 0 \\
 & & \gamma\downarrow & & \downarrow & & \| & & \\
0 & \longrightarrow & \Omega^2 C & \longrightarrow & X_C & \underset{\sigma}{\longrightarrow} & C & \longrightarrow & 0
\end{array}$$

by construction. Given $\rho : D \to C$ we get a diagram

$$\begin{array}{ccccccccc}
0 & \longrightarrow & \Omega D & \overset{i_D}{\longrightarrow} & P_D & \longrightarrow & D & \longrightarrow & 0 \\
 & & \rho'\downarrow & & \downarrow & & \rho\downarrow & & \\
0 & \longrightarrow & \Omega C & \longrightarrow & P_C & \longrightarrow & C & \longrightarrow & 0 \\
 & & \gamma\downarrow & & \downarrow & & \| & & \\
0 & \longrightarrow & \Omega^2 C & \underset{\sigma'}{\longrightarrow} & X_C & \underset{\sigma}{\longrightarrow} & C & \longrightarrow & 0
\end{array}$$

Then

$\rho : D \to C$ is a split epimorphism

$\Leftrightarrow$ $\rho_* : (C,D)^{1,G} \to (C,C)^{1,G}$ has 1_C in its image

$\Leftrightarrow$ ρ_* is surjective

$\Leftrightarrow$ $\rho^\# : \mathrm{Ext}^1_G(C,\Omega^2 C)\ (\cong(\Omega C, \Omega^2 C)^{1,G}) \to \mathrm{Ext}^1_G(D,\Omega^2 C)(\cong(\Omega D,\Omega^2 C)^{1,G})$ is injective

where $\rho^\#$ is the adjoint of ρ_* under the duality given in theorem 2.17.5.

$\Leftrightarrow$ $\rho^{\#}$ does not have γ in its kernel

$\Leftrightarrow$ $\rho'\gamma \notin (\Omega D,\Omega^2 C)_1^G$

$\Leftrightarrow$ $\rho'\gamma$ does not factor through i_D

$\Leftrightarrow$ ρ does not factor through σ .

For if $\lambda\colon P_D \to \Omega^2 C$ with $i_D\lambda = \rho'\gamma$ and $\mu\colon P_D \to P_C \to X_C$ then $\mu - \lambda\sigma'$ vanishes on ΩD, and so gives a map from D to X_C whose composite with σ is ρ.

(ii) We now prove **uniqueness**. Suppose $0 \to A \to B \to C \to 0$ is another almost split sequence terminating in C. Then we get a diagram

$$\begin{array}{ccccccccc}
0 & \longrightarrow & \Omega^2 C & \longrightarrow & X_C & \longrightarrow & C & \longrightarrow & 0 \\
 & & \downarrow{\scriptstyle \alpha_1} & & \downarrow & & \| & & \\
0 & \longrightarrow & A & \longrightarrow & B & \longrightarrow & C & \longrightarrow & 0 \\
 & & \downarrow{\scriptstyle \alpha_2} & & \downarrow & & \| & & \\
0 & \longrightarrow & \Omega^2 C & \longrightarrow & X_C & \longrightarrow & C & \longrightarrow & 0
\end{array} .$$

The map $\alpha_1\alpha_2$ is not nilpotent, and is hence an isomorphism by 1.3.3. Hence by the five-lemma, the two sequences are isomorphic. □

2.17.8 <u>Proposition</u>

If $0 \to A \xrightarrow{\sigma'} B \to C \to 0$ is an almost split sequence and $\rho\colon A \to D$ is not a split monomorphism (i.e. unless A is isomorphic to a direct summand of D and ρ is the injection) then ρ factors through σ'.

<u>Proof</u>

Suppose ρ does not factor through σ'. Then in the pushout

$$\begin{array}{ccccccccc}
0 & \longrightarrow & A & \xrightarrow{\sigma'} & B & \longrightarrow & C & \longrightarrow & 0 \\
 & & \downarrow{\scriptstyle \rho} & & \downarrow & & \| & & \\
0 & \longrightarrow & D & \longrightarrow & E & \longrightarrow & C & \longrightarrow & 0
\end{array}$$

the second sequence does not split. Thus we may complete a diagram

$$\begin{array}{ccccccccc}
0 & \longrightarrow & A & \xrightarrow{\sigma'} & B & \longrightarrow & C & \longrightarrow & 0 \\
 & & \downarrow{\scriptstyle \rho} & & \downarrow & & \| & & \\
0 & \longrightarrow & D & \longrightarrow & E & \longrightarrow & C & \longrightarrow & 0 \\
 & & \downarrow{\scriptstyle \rho'} & & \downarrow & & \| & & \\
0 & \longrightarrow & A & \longrightarrow & B & \longrightarrow & C & \longrightarrow & 0
\end{array}$$

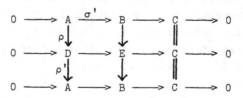

Since $\rho\rho'$ is not nilpotent it is an isomorphism, and so ρ is a split monomorphism. □

2.17.9 <u>Corollary</u>

If $0 \to A \to B \to C \to 0$ is an almost split sequence then so is

$0 \to C^* \to B^* \to A^* \to 0.$ □

2.17.10 Proposition

An almost split sequence $0 \to A \to B \to C \to 0$ splits on restriction to a subgroup H if and only if H does not contain a vertex of C (or equivalently a vertex of A)

Proof

Suppose the sequence splits on restriction to H. Then for any H-module V,

$$\dim_k \mathrm{Hom}_{kH}(V, B{\downarrow_H}) = \dim_k \mathrm{Hom}_{kG}(V, A{\downarrow_H}) + \dim_k \mathrm{Hom}_{kH}(V, C{\downarrow_H})$$

and so by Frobenius reciprocity (2.1.3)

$$\dim_k \mathrm{Hom}_{kG}(V{\uparrow^G}, B) = \dim_k \mathrm{Hom}_{kG}(V{\uparrow^G}, A) + \dim_k \mathrm{Hom}_{kG}(V{\uparrow^G}, C).$$

This means that

$$0 \to \mathrm{Hom}_{kG}(V{\uparrow^G}, A) \to \mathrm{Hom}_{kG}(V{\uparrow^G}, B) \xrightarrow{\phi} \mathrm{Hom}_{kG}(V{\uparrow^G}, C) \to 0$$

is exact (cf. 1.3). Thus C is not a direct summand of $V{\uparrow^G}$, since otherwise by definition of almost split sequence, the projection $V{\uparrow^G} \to C$ would not be in $\mathrm{Im}(\phi)$.

Conversely if the sequence does not split on restriction to H then the identity map on $C{\downarrow_H}$ is not in the image of $\mathrm{Hom}_{kH}(C{\downarrow_H}, B{\downarrow_H}) \to \mathrm{Hom}_{kH}(C{\downarrow_H}, C{\downarrow_H})$ and hence $\mathrm{Fr}'_{H,G}(1_{C{\downarrow_H}})$ is not in the image of $\mathrm{Hom}_{kG}(C{\downarrow_H}{\uparrow^G}, B) \to \mathrm{Hom}_{kG}(C{\downarrow_H}{\uparrow^G}, C)$ (2.1.3). Thus the natural map $\mathrm{Fr}'_{H,G}(1_{C{\downarrow_H}}): C{\downarrow_H}{\uparrow^G} \to C$ does not lift to a map from $C{\downarrow_H}{\uparrow^G}$ to B, and so by definition of almost split sequence, it is a split epimorphism. Thus H contains a vertex of C (2.3.2). □

Remark

In fact, almost split sequences also exist for lattices over an R-order (see [6], [76], [80], [81]) and hence for RG-modules (recall the convention introduced in 1.7 that RG-module means finitely generated R-free RG-module). The construction, however, is very different, so that for kG-modules $A \cong \Omega^2 C$ while for RG-modules $A \cong \Omega C$. In some sense, this corresponds to the fact that for kG-modules $((U,V)^{1,G})^* \cong (V, \Omega U)^{1,G}$ (see 2.17.1 and 2.17.5) while for RG-modules, $((U,V)^G)^* \cong (V,U)^G$.

The existence of almost split sequences for RG-modules does not seem to lead naturally to non-singularity results of the type proved in 2.18, but most of the theory developed in sections 2.28 - 2.32 applies with not much change, to modules for RG (see [92]).

2.18 Non-singularity of the Inner Products on $A(G)$

In this section we use the almost split sequences, constructed in the last section, to investigate the inner products $(\, , \,)$ and $< \, , \, >$ on $A(G)$.

2.18.1 Lemma

If C and D are indecomposable kG-modules, and

$$0 \longrightarrow \Omega^2 C \longrightarrow X_C \longrightarrow C \longrightarrow 0$$

is the almost split sequence terminating in C, then the following hold.

(i) If $C \not\cong D$ then

$$0 \to (D, \Omega^2 C)^G \to (D, X_C)^G \to (D, C)^G \to 0$$

is exact.

(ii) The sequence

$$0 \to (C, \Omega^2 C)^G \to (C, X_C)^G \to (C, C)^G \to \mathrm{SocExt}^1_G(C, \Omega^2 C) \to 0$$

is exact, where this is the truncation of the long exact Ext sequence.

Proof

This follows immediately from the proof of 2.17.7. □

2.18.2 Definition

If V is an indecomposable kG-module, let

$$d_V = \dim_k(\mathrm{End}_{kG}(V)/J(\mathrm{End}_{kG}(V))).$$

Note that if k is algebraically closed then $d_V = 1$ for all modules V.

Let $$\tau(V) = \begin{cases} \mathrm{Soc}(V) & \text{if } V \text{ is projective} \\ X_{\mho V} - \Omega V - \mho V & \text{otherwise} \end{cases}$$

as an element of $A(G)$, where

$$0 \longrightarrow \Omega V \longrightarrow X_{\mho V} \longrightarrow \mho V \longrightarrow 0$$

is the almost split sequence terminating in $\mho V$. Then $\tau(V)$ is called the atom corresponding to V. We extend τ to a semilinear map on $A(G)$ by setting

$$\tau(\Sigma a_i V_i) = \Sigma \, \bar{a}_i \tau(V_i).$$

The reasons for these definitions will become apparent.

2.18.3 Lemma

$$V.\tau(V) = \begin{cases} V - \mathrm{Rad}(V) & \text{if } V \text{ is projective} \\ V + \Omega^2 V - X_V & \text{otherwise} \end{cases}$$

where

$$0 \to \Omega^2 V \to X_V \to V \to 0$$

is the almost split sequence terminating in V.

(Recall $v = P_1 - \Omega(1)$)

Proof

If V is projective, this is 2.4.2(iii). If V is not projective, then 2.4.2(iii) shows that $v.\tau(V) \equiv -\Omega(\tau(V)) = V + \Omega^2 V - X_V$ modulo projectives. But by 2.11.3, $A(G) = A(G,1) \oplus A_0(G,1)$. Since $\tau(V) \in A_0(G,1)$, so is $v.\tau(V)$. Since $V + \Omega^2(V) - X_V$ is also in $A_0(G,1)$, this proves the lemma. □

2.18.4 Theorem

$$(V, v.\tau(W)) = \langle V, \tau(W) \rangle = \begin{cases} d_V & \text{if } V \cong W \\ 0 & \text{otherwise.} \end{cases}$$

Proof

If W is projective, this follows from 2.4.4. Otherwise,

$$\begin{aligned} \langle V, \tau(W) \rangle &= (V, v.\tau(W)) && \text{by 2.4.3} \\ &= (V, W + \Omega^2 W - X_W) && \text{by 2.18.3} \\ &= \begin{cases} d_V & \text{if } V \cong W \\ 0 & \text{otherwise} \end{cases} && \text{by 2.18.1.} \end{aligned}$$ □

2.18.5 Corollary

$\langle\, , \,\rangle$ and $(\, , \,)$ are non-singular on $A(G)$, in the sense that given $x \neq 0$ in $A(G)$, there is a $y \in A(G)$ such that $\langle x,y \rangle \neq 0$ and a $z \in A(G)$ such that $(x,z) \neq 0$.

Proof

If $x = \Sigma a_i V_i$ then $\tau(x) = \Sigma\, \bar{a}_i \tau(V_i)$, and so

$$\langle x, \tau(x) \rangle = \Sigma |a_i|^2 \, d_{V_i} \geq 0$$

with equality if and only if $x = 0$. Thus we may take $y = \tau(x)$ and $z = v.y$. □

2.18.6 Corollary

Suppose U and V are two kG-modules, and for every kG-module X,

$$\dim_k \mathrm{Hom}_{kG}(U,X) = \dim_k \mathrm{Hom}_{kG}(V,X).$$

Then $U \cong V$.

Proof

This follows immediately from 2.18.5. □

2.18.7 Corollary

Suppose $A(G) = A_1 \oplus A_2$ is an ideal direct sum decomposition of

$A(G)$, and suppose A_1 and A_2 are closed under the automorphism $*$ of $A(G)$ given by taking dual modules. Then $\langle\ ,\ \rangle$ and $(\ ,\ )$ are non-singular on A_1 and A_2, and in $A(G)$, $A_1^{\perp} = A_2$ and $A_2^{\perp} = A_1$ with respect to either inner product.

Proof

Let π_1 and π_2 be the projections of $A(G)$ onto A_1 and A_2. Then given $x \in A_1$ and $y \in A_2$, we have

$$\langle x,y\rangle = \langle 1,x^*.y\rangle = \langle 1,0\rangle = 0$$

since $A_1.A_2 = 0$. Thus if $x \neq 0$, $\langle x,\pi_1(\tau(x))\rangle = \langle x,\tau(x)\rangle \neq 0$. Thus $\langle\ ,\ \rangle$ is non-singular on A_1 and $A_1^{\perp} = A_2$. The same argument works for $(\ ,\ )$, with $v.\tau(x)$ in place of $\tau(x)$. □

2.18.8 Corollary

(i) $Im(i_{H,G}) = Ker(r_{G,H})^{\perp}$

(ii) $Ker(r_{G,H}) = Im(i_{H,G})^{\perp}$

(iii) $A(G,H) = A_0(G,H)^{\perp}$

(iv) $A_0(G,H) = A(G,H)^{\perp}$.

Proof

(i) and (ii) follows from 2.18.7 and 2.2.2(i), while (iii) and (iv) follow from 2.18.7 and 2.15.6. □

2.18.9 Corollary

The following are equivalent condition on an indecomposable kG-module V.

(i) V is H-projective.

(ii) $\tau(V)\downarrow_H \neq 0$.

(iii) Tensoring with V splits every H-split short exact sequence.

Proof

(i) ⇔ (ii) follows from 2.17.10, while (i) ⇔ (iii) follows from 2.18.8 (iii). □

2.18.10 Definitions

The glue for a short exact sequence $0 \to X' \to X \to X'' \to 0$ is the element $X - X' - X''$ of $A(G)$. Thus if V is a non-projective indecomposable then $\tau(V)$ is a glue.

A glue is irreducible if it is non-zero, and is not the sum of two non-zero glues as an element of $A(G)$.

2.18.11 Lemma

If $X - X' - X''$ is the glue for $0 \to X' \to X \to X'' \to 0$ then for any module V, $\langle X - X' - X'', V \rangle \geq 0$.

Proof

The number of copies of P_1 in the direct sum decomposition of $X \otimes V$ is at least the sum of the number of copies in $X' \otimes V$ and the number of copies in $X'' \otimes V$ since P_1 is both projective and injective. □

2.18.12 Theorem

(i) Every non-zero glue can be written as the sum of an atom and a glue. Thus every irreducible glue is an atom.

(ii) The atoms are precisely the simple modules and the irreducible glues.

Proof

First we note that the sum of two glues is a glue, since we can add the exact sequences term by term as a direct sum.

(i) Suppose $0 \to Y' \to Y \overset{\pi}{\to} Y'' \to 0$ is an exact sequence with $Y - Y' - Y'' \neq 0$ its glue. If Y'' is decomposable, $Y'' = W'' \oplus Z''$, then $Y - Y' - Y''$ is the sum of the glues for

$$0 \to \pi^{-1}(W'') \to Y \to Z'' \to 0$$

and

$$0 \to Y' \to \pi^{-1}(W'') \to W'' \to 0.$$

At least one of these is non-zero, and so we may assume by induction that Y'' is indecomposable. Thus π is not a split epimorphism. Letting $0 \to \Omega^2(Y'') \to X_{Y''} \to Y'' \to 0$ be the almost split sequence terminating in Y'', we have the following commutative diagram.

$$\begin{array}{ccccccccc} 0 & \longrightarrow & Y' & \longrightarrow & Y & \longrightarrow & Y'' & \longrightarrow & 0 \\ & & \downarrow & & \downarrow & & \downarrow & & \\ 0 & \longrightarrow & \Omega^2 Y'' & \longrightarrow & X_{Y''} & \longrightarrow & Y'' & \longrightarrow & 0 \end{array}$$

The left-hand square is a pushout diagram, and so we get an exact sequence

$$0 \longrightarrow Y' \longrightarrow Y \oplus \Omega^2 Y'' \longrightarrow X_{Y''} \longrightarrow 0 .$$

The given glue is the sum of the glue for this sequence and the atom corresponding to the almost split sequence terminating in Y''.

(ii) If V is a projective indecomposable module, then $\tau(V)$ is a simple module. If V is a non-projective indecomposable, then $\tau(V)$ is a glue. Suppose it is not irreducible. Then by (i) it is the sum of another atom, say $\tau(W)$, and a glue. But $< \tau(V) - \tau(W), W > = -d_W$ by 2.18.4, contradicting 2.18.11. Conversely by (i), every irreducible glue is an atom, and clearly every simple module is an atom $(V = \tau(P_V))$. □

If k is algebraically closed, we formally think of each representation V as consisting of (possibly infinitely many) atoms, namely the simple composition factors and some irreducible glues holding them together.

$$V = \sum_{\substack{W \\ \text{indec.}}} < V,W > \tau(W)$$

This formal expression has the right inner product with each indecomposable module Y, because each d_Y is one, and so

$$\begin{aligned} < (\Sigma < V,W > \tau(W)), Y > &= \Sigma << V,W > \tau(W), Y > \\ &= < V,Y > \text{ by } 2.18.4. \end{aligned}$$

Thus the expression has the right inner product with any element of $A(G)$, and so since the inner products are non-singular, this is a reasonable formal sum to write down.

We consider atoms to be in the same block as the corresponding indecomposable modules. Then in the formal sum above, an indecomposable module can only involve atoms from the same block.

Exercises

1. Suppose k_1 is an extension of k. Show that the natural map $A_k(G) \to A_{k_1}(G)$ preserves the inner products $(\ , \ )$ and $< \ , \ >$. Use 2.18.5 to deduce that this map is injective. This is called the Noether-Deuring theorem, and a more conventional proof is given in [37], p. 200-202.

Now suppose k_1 is a separable algebraic extension of k. Show that $A_{k_1}(G)$ is integral as an extension of $A_k(G)$.

2. (i) Show that a short exact sequence $0 \to X' \to X \to X'' \to 0$ splits if and only if the glue $X - X' - X''$ is zero, namely if and only if $X \cong X' \oplus X''$ (hint: examine the long exact sequence

associated with $Hom_{kG}(X'',-)$ and count dimensions).

(ii) Show that a sequence $0 \to \Omega^2 U \to X \to U \to 0$ (with U indecomposable) is almost split if and only if its glue is irreducible.

(iii) Trying to generalize (i) and (ii), we might conjecture that whenever we have two short exact sequences $0 \to X_1' \to X_1 \to X_1'' \to 0$ and $0 \to X_2' \to X_2 \to X_2'' \to 0$ with $X_1' \cong X_2'$, $X_1 \cong X_2$ and $X_1'' \cong X_2''$, there is an isomorphism of short exact sequences. The following is a counter-example. Let G be the fours group (a direct product of two copies of the cyclic group of order two), and k a field of characteristic two. Let $X_1 = X_2 = \Omega^2(k)$. Then there are isomorphic one dimensional submodules X_1' and X_2' with $X_1/X_1' \cong X_2/X_2' \cong k \oplus \Omega(k)$, but there is no automorphism of X_1 taking X_1' to X_2'.

3. (i) Show that there is an almost split sequence

$$0 \to \Omega^2(k) \to \Omega(Rad(P_1)/Soc(P_1)) \to k \to 0$$

(ii) Let U and V be indecomposable modules and suppose k is <u>algebraically closed.</u> Show that $U^* \otimes V$ has the trivial module as a direct summand if and only if the following two conditions are satisfied

(a) $U \cong V$ and

(b) $p \nmid dim(U)$

(hint: consider the composite map $Hom_{kG}(U,V) \to U \otimes V^* \to (Hom_{kG}(V,U))^*$; the corresponding map $Hom_{kG}(U,V) \otimes Hom_{kG}(V,U) \to k$ is given by $a \otimes b \to Tr(ab)$. This factors as $Hom_{kG}(U,V) \otimes Hom_{kG}(V,U) \to End_{kG}(U) \to k$. Now use 1.3.3.)

(iii) Let U be an indecomposable kG-module. Tensoring the sequence of (i) with U and using the fact that $\Omega^2(k) \otimes U \cong \Omega^2(U) \oplus$ projective (Schanuel's lemma), we obtain a short exact sequence

$$0 \to \Omega^2(U) \to X \to U \to 0.$$

Show that this sequence is always either split or almost split, and is almost split if and only if $p \nmid dim(U)$. (Hint: let $x = k + \Omega^2(k) - \Omega(Rad(P_1)/Soc(P_1))$; use the identity $(U^*.V,x)=(V,U.x)$, together with (ii) and question 2).

(iv) Show that the linear span $A(G;p)$ in $A(G)$ of the indecomposable modules whose dimension is divisible by p is an ideal, and that $A(G)/A(G;p)$ **has no non-zero nilpotent elements.**

(v) Suppose H is p-hypoelementary with $O_p(H)$ cyclic. Show that $A(H;p)$ consists of induced modules. Use 2.2.2 and induction to show that $A(H)$ is semisimple.

(vi) Using the results of 2.13, show that for any group G,

A(G,Cyc), the linear span of the modules whose vertex is cyclic, is semisimple.

4. Let G be an elementary abelian group of order eight, k a field of characteristic two, and $V = \mathrm{Rad}(P_1)/\mathrm{Soc}(P_1)$.

(i) Show that if W is a submodule of V then either $\dim(W)=1$ or $\dim(W \cap \mathrm{Soc}(V)) > 1$. Deduce that V is indecomposable.

(ii) Using 3(iii), show that $V \otimes V \cong \Omega(V) \oplus \mho(V)$. Thus V is self-dual, but $V \otimes V$ has no self-dual indecomposable summands.

2.19 The Radical of $\dim_k \mathrm{Ext}^n_G$

We define bilinear forms $(\ ,\ )_n$ for $n \geq 1$ as follows. If U and V are kG-modules, we let

$$(U,V)_n = \dim_k \mathrm{Ext}^n_G(U,V).$$

We extend this bilinearly to give (not necessarily symmetric) bilinear forms on the whole of A(G). The purpose of this section is to use the results of the last section to obtain information about the radicals of these forms, which in fact turn out all to be the same.

2.19.1 Lemma

(i) There is a natural isomorphism $\mathrm{Ext}^n_G(U,V) \cong (\Omega^n U,V)^{1,G}$

(ii) $(U,V)_n = (\Omega^n U,V) - \langle \Omega^n U,V \rangle = ((1-u)\Omega^n U,V)$.

Proof

(i) The short exact sequence

$$0 \to \Omega U \to P_U \to U \to 0$$

gives rise to a long exact sequence

$$\begin{aligned} 0 &\to (U,V)^G \to (P_U,V)^G \xrightarrow{\sigma} (\Omega U,V)^G \to \\ &\to \mathrm{Ext}^1_G(U,V) \to \underset{=0}{\mathrm{Ext}^1_G(P_U,V)} \to \mathrm{Ext}^1_G(\Omega U,V) \to \\ &\to \mathrm{Ext}^2_G(U,V) \to \underset{=0}{\mathrm{Ext}^2_G(P_U,V)} \to \cdots \end{aligned}$$

Thus $\mathrm{Ext}^1_G(U,V) \cong (\Omega U,V)^G/\mathrm{Im}(\sigma) = (\Omega U,V)^{1,G}$, and for $n \geq 2$, $\mathrm{Ext}^n_G(U,V) \cong \mathrm{Ext}^{n-1}_G(\Omega U,V)$.

(ii) $\dim(\Omega^n U,V)^{1,G} = (\Omega^n U,V) - \langle \Omega^n U,V \rangle$ by (i)
$= ((1-u)\Omega^n U,V)$ by 2.4.3. □

2.19.2 Definition

$$\mathrm{Rad}(\ ,\ )_n = \{x \in A(G) : (x,y)_n = 0 \text{ for all } y \in A(G)\}.$$

Since 2.4.3 implies that $(x,y) = (u^2y,x)$ we have

$$\mathrm{Rad}(\ ,\ )_n = \{x \in A(G) : (y,x)_n = 0 \text{ for all } y \in A(G)\}.$$

2.19.3 Lemma

Suppose U is a periodic kG-module with even period $2s$, i.e. $\Omega^{2s}(U) \cong U$. Then as elements of $A(G)$,

$$U = u^{2s}.U.$$

Proof

By 2.4.2(ii), we have the following congruences modulo $A(G,1)$.

$$U \equiv -u.\Omega(U) \equiv u^2.\Omega^2(U) \equiv \ldots \equiv u^{2s}.\Omega^{2s}U = u^{2s}.U\ .$$

Now $u \equiv 1$ modulo $A_0(G,1)$, and so $U \equiv u^{2s}.U$ modulo $A_0(G,1)$. The result now follows from 2.11.3 □

2.19.4 Theorem

$\mathrm{Rad}(\ ,\ )_n$ is the linear span in $A(G)$ of the projective modules and elements of the form

$$\sum_{i=1}^{2s} (-1)^i\ \Omega^i(U)$$

for U a periodic module of even period $2s$.

Proof

Suppose $x = \Sigma a_i V_i \in \mathrm{Rad}(\ ,\ )_n$. Then for V_i non-projective we have

$$\begin{aligned} 0 &= (x,\tau(\Omega^n V_i))_n \\ &= (\Omega^n x,\tau(\Omega^n V_i)) - <\Omega^n x,\tau(\Omega^n V_i)> \quad \text{by 2.19.1} \\ &= (x,\ \tau(V_i)) - <x,\tau(V_i)> \\ &= -\ (x,v.\tau(\mho V_i)) - <x,\tau(V_i)> \end{aligned}$$

(since by 2.18.3 if V_i is non-projective $\tau(V_i) = -v.\tau(\mho V_i)$)

$$\begin{aligned} &= -\ <x,\tau(\mho V_i)> - <x,\tau(V_i)> \\ &= -\ (\text{coefficient of } \mho V_i) - (\text{coefficient of } V_i) \end{aligned}$$

Hence

$$(\text{coefficient of } V_i) = -\ (\text{coefficient of } \Omega V_i)$$

Thus if $a_i \neq 0$, V_i is projective or periodic of even period.

Conversely, if V is periodic of even period $2s$, then by 2.19.3 we have

$$(1-u)(1+u+\ \ldots\ +u^{2s-1})V = 0$$

and so $(1+u+ \dots +u^{2s-1})V \in \mathrm{Rad}(\ ,\ )_n$ by 2.19.1. But $(1+u+ \dots + u^{2s-1})V \equiv \sum_{i=1}^{2s} (-1)^i \Omega^i V$ modulo projectives by 2.4.2(iii). □

2.20 The Atom Copying Theorem

This is a very short section in which we use the Burry-Carlson Theorem to investigate the behaviour of atoms under induction. For simplicity we assume k is algebraically closed.

2.20.1 Theorem (Atom Copying by Induction)

Let D be a p-subgroup of G, and let H be a subgroup of G with $N_G(D) \leq H$. Let V be an indecomposable kG-module with vertex D and Green correspondent W. Denote by τ the map given in 2.18.2 both for G and for H. Then

$$\tau(W)\uparrow^G = \tau(V) .$$

Proof

By the Burry-Carlson theorem (2.12.3), if U is an indecomposable kG-module, then $U\downarrow_H$ has W as a direct summand if and only if $U \cong V$, and then only once. Hence

$$\langle U,\tau(W)\uparrow^G - \tau(V) \rangle = \langle U\downarrow_H,\tau(W)\rangle - \langle U,\tau(V) \rangle \quad \text{by 2.4.6}$$
$$= 0 \quad \text{by 2.18.4},$$

since all the d_V's are 1. Hence by 2.18.5 $\tau(W)\uparrow^G - \tau(V) = 0$. □

Exercise

Suppose V is a kG-module with a Sylow p-subgroup P as vertex, and W is the Green correspondent of V as a $kN_G(P)$-module. Show that $\tau(V)\downarrow_{N_G(P)} = \tau(W)$.

2.21 The Discrete Spectrum of A(G)

In this section, we investigate what happens when we project the information we have obtained onto a finite dimensional direct summand of $A(G)$ satisfying certain natural conditions (2.21.1). We obtain a pair of dual tables T_{ij} and U_{ij} analogous to the tables of values of Brauer species on the set of irreducible modules and the set of projective indecomposable modules. Indeed, the 'Brauer summand' $A(G,1)$

turns out to be the unique minimal case (2.21.9). The set of species of all such summands forms the 'discrete spectrum' of $A(G)$.

2.21.1 Hypothesis

$A(G) = A \oplus B$ is an ideal direct sum decomposition, with projections $\pi_1 : A(G) \to A$ and $\pi_2 : A(G) \to B$. The summand A satisfies the following conditions.

(i) A is finite dimensional

(ii) A is semisimple as a ring

(iii) A is freely spanned as a vector space by indecomposable modules

(iv) A is closed under taking dual modules.

Remarks

(i) Any finite dimensional semisimple ideal I is a direct summand, since

$$A(G) = I \oplus \bigcap_{s} \mathrm{Ker}(s)$$

where s runs over the set of species of $A(G)$ not vanishing on I. (Note that if I as an ideal of $A(G)$ then any species s of I extends uniquely to a species of $A(G)$. For let $x \in I$ with $(s,x) = 1$. Then for any $y \in A(G)$, and any extension t of s to $A(G)$, we have $(t,y) = (t,y)(s,x) = (t,y)(t,x) = (t,xy) = (s,xy)$. Moreover, it is easy to check that $(t,y) = (s,xy)$ does indeed define a species of $A(G)$.)

(ii) If A satisfies 2.21.1 (i), (ii) and (iii) then the span in $A(G)$ of A and the duals of modules in A form a summand satisfying (i), (ii), (iii) and (iv). Thus (iv) is not a very severe restriction.

If A_1 and A_2 are summands both satisfying 2.21.1 then so are $A_1 + A_2$ and $A_1 \cap A_2$. We define $A(G,\mathrm{Discrete})$ to be the sum of all A satisfying 2.21.1. Any element of $A(G,\mathrm{Discrete})$ lies in some summand A satisfying 2.21.1.

We write $A_0(G,\mathrm{Discrete})$ for the intersection of the B's given in 2.21.1. Note that $A(G,\mathrm{Discrete}) \oplus A_0(G,\mathrm{Discrete})$ is not necessarily the whole of $A(G)$ (the fours group is a counterexample).

2.21.2 Conjecture

Let $H \le G$. Then

(i) $r_{G,H}(A(G,\mathrm{Discrete})) \subseteq A(H,\mathrm{Discrete})$

(ii) $i_{H,G}(A(H,\mathrm{Discrete})) \subseteq A(G,\mathrm{Discrete})$

2.21.3 Examples

(i) By 2.11.3, $A = A(G,1)$, $B = A_o(G,1)$ satisfy 2.21.1. We shall call this case the Brauer case.

(ii) It is shown in [M.F. O'Reilly, 'On the Semisimplicity of the modular representation algebra of a finite group', Ill. J. Math. 9 (1965), 261-276] that the ideal $A(G,Cyc)$, spanned by all the $A(G,H)$ for H cyclic, is a finite dimensional semisimple ideal (see also exercise 3 to 2.18). We write $A(G) = A(G,Cyc) \oplus A_o(G,Cyc)$. Thus $A = A(G,Cyc)$ and $B = A_o(G,Cyc)$ satisfy 2.21.1. We shall call this case the cyclic vertex case, since $A(G,Cyc)$ has a basis consisting of the modules with cyclic vertex.

(iii) Let G be the Klein fours group and k an algebraically closed field of characteristic 2. Then $A_k(G)$ has infinitely many summands satisfying 2.21.1. Thus $A(G,Discrete)$ is infinite dimensional. It turns out that $A_o(G,Discrete)$ is isomorphic to the ideal of $\mathbb{C}[X,X^{-1}]$ consisting of those functions which vanish at $X = 1$ (for more information see the appendix). Thus the set of species of $A(G)$ breaks up naturally into a discrete part and a continuous part. Is there a general theorem along these lines?

2.21.4 Lemma

Suppose $A(G) = A \oplus B$ as in hypothesis 2.21.1. Then $\langle\ ,\ \rangle$ and $(\ ,\ )$ are non-singular on A.

Proof

This is a special case of 2.18.7. □

2.21.5 Definitions

Let $s_1, \ldots, s_n$ be the species of A, and $V_1, \ldots, V_n$ the indecomposable modules freely spanning A. Let $G_i = \tau(V_i)$ (see 2.18.2).

The atom table of A is the matrix

$$T_{ij} = (s_j, G_i) = (s_j, \pi_1(G_i)) .$$

The representation table of A is the matrix

$$U_{ij} = (s_j, V_i) .$$

Let $\Lambda = \pi_1(a(G))$ and $\Lambda_o = A \cap a(G)$.

2.21.6 Lemma

Λ and Λ_o are lattices in A, and $|\Lambda/\Lambda_o| = \det(\langle V_i, V_j\rangle)$ if k is algebraically closed.

Proof

For $x \in a(G)$, $\langle\pi_1(x), V_i\rangle = \langle x, V_i\rangle \in \mathbb{Z}$. Moreover, $\langle G_j, V_i\rangle = \delta_{ij}$ if k is algebraically closed. □

2.21.7 Lemma

If $x \in a(G)$ then (s_1,x) is an algebraic integer.

Proof

The $\mathbb{Z}$-span in A of the tensor powers of $\pi_1(x)$ form a sub-lattice of Λ. Since this lattice satisfies A.C.C., this implies that for some m,

$$(\pi_1(x))^m \in \mathbb{Z}\text{-span}(1,\pi_1(x), \ldots , (\pi_1(x))^{m-1}).$$

This gives a monic equation with integer coefficients satisfied by the value of every species of A on x. □

2.21.8 Open Questions

(i) For $x \in a(G)$, is (s_1,x) always a cyclotomic integer?

(ii) Is Λ/Λ_0 always a p-torsion group?

2.21.9 Lemma

$$A(G,1) \subseteq A \qquad (!)$$

Proof

Since $\pi_1(1)$ is the identity element of A, it is non-zero, and hence by 2.21.4, for some j, $\langle V_j,1\rangle = \langle V_j,\pi_1(1)\rangle \neq 0$. Thus by 2.18.4, some V_j is equal to P_1.

Now look at the set of values (b_g,P_1) of Brauer species on P_1. Suppose there are m different values $(b_{g_1},P_1), \ldots, (b_{g_m},P_1)$.

Let N be the kernel of the action of G on P_1. This has order prime to p since P_1 is projective. (In fact $N = O_{p'}(G)$ but we shall not need to know that). Then the b_g's for which $(b_g,P_1) = \dim(P_1)$ are precisely those b_g with $g \in N$ (how can $\dim(P_1)$ be written as a sum of $\dim(P_1)$ roots of unity?). Since the Vandermonde matrix $(b_{g_i},\otimes^j(P_1))$ is non-singular, some polynomial in P_1 (which is hence an element of A) has value $|G/N|$ on those b_g for which $(b_g,P_1) = \dim(P_1)$ and zero on the rest. By 2.11.3, this element must be the group algebra of G/N. Now since A is closed under taking direct summands, every projective module for G/N lies in A. But the idempotent generator e for $A(G/N,1)$ lies in $A(G,1)$ since N has order prime to p, and every Brauer species of G has value 1 on e. Thus e is the idempotent generator of $A(G,1)$, and so $A(G,1) \subseteq A$. □

Since $P_1 \in A$, we may choose our notation so that $P_1 = V_1$. By 2.2.1, the matrix U_{ij} is invertible. We define

$$m_i = (U^{-1})_{i1} = \sum_j (U^{-1})_{ij} \langle 1,V_j\rangle .$$

Thus $<1,V_i> = \sum_j U_{ij}m_j$, and so for any $x \in A(G)$ and $y \in A$ we have the following equations.

$$<1,y> = \sum_j (s_j,y)m_j$$
$$<x,y> = <1,x^*y> = \sum_j (s_j,x^*)(s_j,y)m_j .$$

But now by 2.21.4, this means the m_j are non-zero, and so we may define

$$c_j = c(s_j) = c_G(s_j) = 1/m_j .$$

Thus we have, for $x \in A(G)$, $y \in A$,

2.21.10 $$<x,y> = \sum_j \frac{(s_j,x^*)(s_j,y)}{c_j} .$$

Now let $p_i = (s_i,u) = (s_i,\pi_1(u))$. By 2.4.3, for $x \in A(G)$ and $y \in A$, we have $(x,y) = <x,u.y>$, and so by 2.21.10 we have

2.21.11 $$(x,y) = \sum_j \frac{p_j(s_j,x^*)(s_j,y)}{c_j} .$$

Now let $U^\#$ be the matrix obtained from U by transposing and replacing each representation by its dual. Let C be the diagonal matrix of c_i's. The orthogonality relations 2.21.10 can be written in the form $TC^{-1}U^\# = 1$, i.e. $U^\#T = C$.

2.21.12 <u>Question</u>

Is it true in general that $U^\# = U^\dagger$, the Hermitian adjoint of U? In other words, is it true that $(s,x^*) = \overline{(s,x)}$? This would imply that the c_j are real algebraic numbers. Are they algebraic integers?

2.21.13 <u>Proposition</u>

Suppose $H \le G$, A satisfies 2.21.1, and $r_{G,H}(A) \subseteq A'$ with A' satisfying 2.21.1 (e.g. in the examples of 2.21.3 (i) and (ii), we could let A' be $A(H,1)$ and $A(H,Cyc)$ respectively) Suppose s is a species of A which factors through H, and t is a species of A' fusing to s. Then

$$c_G(s) = |N_G(\mathrm{Orig}(t)) \cap \mathrm{Stab}_G(t) : N_H(\mathrm{Orig}(t))|.c_H(t)$$

<u>Proof</u>

Choose an element $x \in A$ such that s has value 1 on x and all other species of A have value zero on x, and an element $y \in A'$ such that t has value 1 on y^* and all other species of A' have value zero on y^*. Then by 2.4.6, $<y\uparrow^G,x> = <y,x\downarrow_H>$, and so

$$\sum_i \frac{(s_i, y^*\!\uparrow^G)(s_i, x)}{c_G(s_i)} = \sum_j \frac{(t_j, y^*)(t_j, x\!\downarrow_H)}{c_H(t_j)} .$$

Thus by the choice of x and y,

$$\frac{(s, y^*\!\uparrow^G)}{c_G(s)} = \frac{(t, x\!\downarrow_H)}{c_H(t)}$$

$$= \frac{(s,x)}{c_H(t)} = \frac{1}{c_H(t)} .$$

But the induction formula 2.10.2 gives

$$(s, y^*\!\uparrow^G) = |N_G(\mathrm{Orig}(t)) \cap \mathrm{Stab}_G(t) : N_H(\mathrm{Orig}(t))|$$

thus proving the desired formula. □

2.21.14 <u>Corollary</u>

Suppose s is a species of two different summands A_1 and A_2 of $A(G)$ both satisfying 2.21.1. Then the two definitions of $c_G(s)$ coincide.

<u>Proof</u>

Take $G = H$, $A = A_1$ and $A' = A_1 + A_2$ in the proposition to conclude that the values of $c_G(s)$ as species of A_1 and of $A_1 + A_2$ coincide. □

2.21.15 <u>Corollary</u>

Let $H \le G$ and V be an H-module, and A, A' as in 2.21.13. Then

$$(s, V\!\uparrow^G) = \sum_{s_o \sim s} \frac{c_G(s)}{c_H(s_o)} \quad (s_o, V)$$

where s_o runs over those species of A' fusing to s.

<u>Proof</u>

This follows from 2.21.13 and the induction formula 2.10.2. □

2.21.16 <u>Corollary</u>

Let $A = A(G,1)$. Then $c_G(b_g) = |C_G(g)|$.

<u>Proof</u>

If $G = \langle g \rangle$ this is an easy exercise. For the general case apply 2.21.13 with $H = \langle g \rangle$, $s = b_g$ and $t = b_g$. Then $N_G(\mathrm{Orig}(t)) \cap \mathrm{Stab}_G(t) = C_G(g)$, and $N_H(\mathrm{Orig}(t)) = \langle g \rangle$. Thus

$$c_G(b_g) = |C_G(g) : \langle g \rangle| \cdot |\langle g \rangle| = |C_G(g)|. \quad □$$

2.22 Group Cohomology and the Lyndon-Hochschild-Serre Spectral Sequence $H^p(G/N,H^q(N,V)) \Rightarrow H^{p+q}(G,V)$.

In this chapter I attempt to provide a brief description of the tools from group cohomology theory necessary for the study of complexity theory and varieties for modules (see sections 2.24-2.27). I have made no attempt at completeness.

A free resolution of $\mathbb{Z}$ as a $\mathbb{Z}G$-module is an exact sequence

$$0 \longleftarrow \mathbb{Z} \xleftarrow{\varepsilon} X_0 \xleftarrow{\partial_1} X_1 \xleftarrow{\partial_2} \cdots$$

with the X_i free $\mathbb{Z}G$-modules. The map ε is called the augmentation map. It is easy to see that given two free resolutions we can find maps

$$\begin{array}{ccccccccc} 0 \longleftarrow & \mathbb{Z} & \xleftarrow{\varepsilon} & X_0 & \xleftarrow{\partial_1} & X_1 & \xleftarrow{\partial_2} & X_2 & \longleftarrow \cdots \\ & \| & & \downarrow \lambda_0 & & \downarrow \lambda_1 & & \downarrow \lambda_2 & \\ 0 \longleftarrow & \mathbb{Z} & \xleftarrow{\varepsilon'} & X'_0 & \xleftarrow{\partial'_1} & X'_1 & \xleftarrow{\partial'_2} & X'_2 & \longleftarrow \cdots \end{array}$$

and that any two such sets of maps are chain homotopic, namely given λ_i and λ'_i there are maps $h_i : X_i \to X'_{i+1}$ such that $\lambda_i - \lambda'_i = \partial_i h_{i-1} + h_i \partial'_{i+1}$ (in fact this depends only on the X_i being free).

If V is a $\mathbb{Z}G$-module, taking homomorphisms from (X_i,∂_i) to V gives a cochain complex

$$V^0 \xrightarrow{\delta^0} V^1 \xrightarrow{\delta^1} V^2 \xrightarrow{\delta^2} \cdots$$

with

$$V^i = \mathrm{Hom}_{\mathbb{Z}G}(X_i,V)$$

$$\delta^i = \mathrm{Hom}_{\mathbb{Z}G}(\partial_{i+1},V) \quad \text{(i.e. the map obtained by composition with } \partial_{i+1}\text{)}.$$

We define cohomology groups

$$H^i(X,V) = \mathrm{Ker}(\delta^i)/\mathrm{Im}(\delta^{i+1}) \qquad i > 0$$

$$H^0(X,V) = \mathrm{Ker}(\delta^0) \cong V^G .$$

Given two free resolutions X_i and X'_i, a map of resolutions λ_i gives rise to a map $\lambda_i^*: H^i(X',V) \to H^i(X,V)$, and since any two λ_i are homotopic, they give rise to the same λ_i^*. In particular, if $\mu_i : X'_i \to X_i$ then $\lambda_i\mu_i$ is homotopic to the identity map, and so $\mu_i^*\lambda_i^* = 1$. Thus we have a natural isomorphism $H^i(X,V) \cong H^i(X',V)$, and so the cohomology groups are independent of choice of free resolution.

Thus we may simply write $H^i(G,V)$.

The <u>bar resolution</u> is the free resolution given by letting X_i be the free $\mathbb{Z}G$-module on symbols $[g_1| \dots |g_i]$, $g_j \in G$, and

$$[g_1| \dots |g_n]\partial_n = [g_1| \dots |g_{n-1}]g_n + \sum_{i=1}^{n-1} (-1)^i [g_1| \dots |g_{n-i}g_{n-i+1}| \dots |g_n] + (-1)^n [g_2| \dots |g_n]$$

$$[\,]\varepsilon = 1.$$

The submodules of X_i generated by those $[g_1| \dots |g_n]$ with some $g_i = 1$ form a free subcomplex which we may quotient out to obtain the <u>normalized bar resolution</u> $(\tilde{X}_i(G),\tilde{\partial}_i)$.

<u>Remark</u>

The bar resolution becomes more transparent if we write it in terms of the $\mathbb{Z}$-basis

$$(g_0,\dots,g_n) = [g_0g_1^{-1}|g_1g_2^{-1}|\dots|g_{n-1}g_n^{-1}]g_n$$

$$(g_0,\dots,g_n)g = (g_0g,\dots,g_ng)$$

$$(g_0,\dots,g_n)\partial_n = \sum_{i=0}^{n} (-1)^{n-i}(g_0,\dots,g_{i-1},g_{i+1},\dots,g_n)$$

In particular, this makes it easier to check that it is indeed a resolution.

If X_i and Y_i are resolutions of $\mathbb{Z}$, then so is $(X \otimes Y)_i = \sum_{p+q=i} X_p \otimes Y_q$, with boundary homomorphism

$$(*) \qquad (x \otimes y)\partial = x \otimes y\partial + (-1)^{\deg(y)}\, x\partial \otimes y.$$

Thus there is a map of resolutions $\Delta\colon X_i \to (X \otimes X)_i$, and any two such are homotopic. Such a map is called a <u>diagonal approximation</u>. For the (normalized) bar resolution, we use a particular diagonal approximation called the <u>Alexander-Whitney map</u>

$$[g_1| \dots |g_n]\Delta = \sum_{j=0}^{n} [g_1| \dots |g_j]g_{j+1}\cdots g_n \otimes [g_{j+1}| \dots |g_n].$$

A diagonal approximation gives rise to a cup product on the cochain level as follows. If $f_1 \in U^i$ and $f_2 \in V^j$ then $f_1 \cup f_2 \in (U \otimes V)^{i+j}$ is given by $x(f_1 \cup f_2) = (x\Delta)(f_1 \otimes f_2)$. For example, the Alexander Whitney map gives

$$[g_1| \dots |g_{i+j}](f_1 \cup f_2) = [g_1| \dots |g_i]g_{i+1}\cdots g_{i+j}f_1 \otimes [g_{i+1}| \dots |g_{i+j}]f_2.$$

Since $(f_1 \cup f_2)\delta = f_1 \cup f_2\delta + (-1)^{\deg(f_2)} f_1\delta \cup f_2$ by (*), the cup product of two cocycles is a cocycle, and the cup product of a cocycle and a coboundary, either way round, is a coboundary. Thus we get a cup product structure

$$\cup : H^i(G,U) \otimes H^j(G,V) \to H^{i+j}(G,U \otimes V),$$

and it is easily checked that since any two diagonal approximations are homotopic, all diagonal approximations give rise to the same cup product structure at the level of cohomology. The following properties of the cup product are easy to verify.

2.22.1 Lemma

(i) $(x \cup y) \cup z = x \cup (y \cup z)$

(ii) Let $t : U \otimes V \to V \otimes U$ be the natural isomorphism. Then

$$(x \cup y)t^* = (-1)^{\deg(x)\deg(y)} y \cup x. \quad \square$$

Remark

We shall often denote cup product operations simply by juxtaposition.

The main properties of the cohomology groups $H^i(G,V)$ are as follows.

(i) A short exact sequence $0 \to V_1 \to V_2 \to V_3 \to 0$ gives rise to a long exact sequence in cohomology:

$$0 \to H^0(G,V_1) \to H^0(G,V_2) \to H^0(G,V_3) \to H^1(G,V_1) \to H^1(G,V_2) \to ..$$

(ii) Universal Coefficient Theorem

Suppose G acts trivially on V. Then there is a short exact sequence

$$0 \to H^n(G,\mathbb{Z}) \otimes V \to H^n(G,V) \to \mathrm{Tor}_1^{\mathbb{Z}}(H^{n+1}(G,\mathbb{Z}),V) \to 0$$

which splits, but not naturally.

(iii) A map of groups $G_1 \to G_2$ gives rise to a map of cohomology $H^i(G_2,V) \to H^i(G_1,V)$ for each i, and these maps commute with cup products.

(iv) Künneth Formula

Suppose G acts trivially on V. Then there is a short exact sequence

$$0 \to \sum_{i+j=n} H^i(G_1,\mathbb{Z}) \otimes H^j(G_2,V) \to H^n(G_1 \times G_2,V) \to \sum_{i+j=n+1} \mathrm{Tor}_1^{\mathbb{Z}}(H^i(G_1,\mathbb{Z}),H^j(G_2,V)) \to 0$$

which also splits, but not naturally.

If k is a field, then

$$H^*(G_1 \times G_2,k) \cong H^*(G_1,k) \otimes H^*(G_2,k) \quad \text{as graded rings,}$$

and if U and V are kG_1- and kG_2 -modules

$$H^*(G_1 \times G_2,U \otimes V) \cong H^*(G_1,U) \otimes H^*(G_2,V).$$

Note that the Künneth Formula depends on G_1 and G_2 being finite, whereas most of this section does not.

(v) If V is an kG-module, we may regard V as a $\mathbb{Z}G$-module. Then V^i, and hence $H^i(G,V)$, have natural k-module structures, and for $i > 0$, $H^i(G,V) \cong \mathrm{Ext}^i_G(k,V)$.

(vi) Suppose G_1 is a subgroup of the group of units of kG, whose elements are of the form $\Sigma a_i g_i$ with $\Sigma a_i = 1$. Then since the concepts of resolution and diagonal approximation are purely module theoretic constructions, we get a homomorphism of cohomology $H^i(G,V) \to H^i(G_1,V)$ commuting with cup products, in the sense that

$$\begin{array}{ccc} H^i(G,k) \otimes H^j(G,V) & \xrightarrow{\cup} & H^{i+j}(G,V) \\ \downarrow & & \downarrow \\ H^i(G_1,k) \otimes H^j(G_1,V) & \xrightarrow{\cup} & H^{i+j}(G_1,V) \end{array}$$

commutes. Beware that this makes no sense with k replaced by an arbitrary module U, since the action of kG on $U \otimes V$ does not commute with the inclusion $kG_1 \hookrightarrow kG$.

(vii) In [48] it is shown that if Λ is a commutative ring satisfying A.C.C. then $H^*(G,\Lambda)$ (regarding Λ as a $\mathbb{Z}G$-module with trivial action) satisfies A.C.C., and that if V is a ΛG-module finitely generated over Λ, then $H^*(G,V)$ is a finitely generated module for $H^*(G,\Lambda)$. By 2.21.1, $H^*(G,\Lambda)$ is not necessarily commutative, but the subring $H^{ev}(G,\Lambda) = \oplus H^{2i}(G,\Lambda)$ is commutative.

Now suppose V is a kG-module. We form the Poincaré series

$$\xi_V(t) = \Sigma\, t^n \dim_k(H^n(G,V)).$$

Then since $H^*(G,V)$ is a finitely generated module over $H^*(G,k)$, it follows from 1.8.2 and the remark following it that $\xi_V(t)$ is a rational function of the form $f(t)/\prod_{i=1}^{r}(1-t^{k_i})$ where $k_1, \ldots, k_r$ are the degrees of a set of homogeneous generators of $H^*(G,k)$, and $f(t)$ is a polynomial with integer coefficients.

(viii) If H is a subgroup of G, we have natural maps

$$\mathrm{res}_{G,H}: H^*(G,V) \to H^*(H,V{\downarrow_H})$$
$$\mathrm{tr}_{H,G}: H^*(H,V{\downarrow_H}) \to H^*(G,V)$$
$$\mathrm{norm}_{H,G}: H^*(H,V) \to H^*(G,V^{\otimes\uparrow G})$$

given as follows. A resolution of $\mathbb{Z}$ as a $\mathbb{Z}G$-module is also a resolution as a $\mathbb{Z}H$-module, and $\mathrm{Hom}_{\mathbb{Z}G}(X_i,V) \subseteq \mathrm{Hom}_{\mathbb{Z}H}(X_i,V)$. Thus G-cocycles may be regarded as H-cocycles, and G-coboundaries are H-coboundaries. Thus we obtain a (not necessarily injective) map $\mathrm{res}_{G,H}:H^i(G,V) \to H^i(H,V{\downarrow}_H)$ (this is the map induced by the inclusion $H \hookrightarrow G$). Similarly, let $\{g_j\}$ be a set of right coset representatives of H in G. Then $\mathrm{Tr}_{H,G}(x) = \Sigma\, g_j^{-1}xg_j$ is in $\mathrm{Hom}_{\mathbb{Z}G}(X_i,V)$, and $\mathrm{Tr}_{H,G}$ commutes with δ_i. Thus we obtain a map $\mathrm{tr}_{H,G}:H^i(H,V{\downarrow}_H) \to H^i(G,V)$.

Now if $X = \{X_i,\partial_i,\varepsilon\}$ is a free resolution of V as a $\mathbb{Z}H$-module, then we can form the tensor product of chain complexes $\bigotimes_j (X \otimes g_j)$ as a complex of $\mathbb{Z}G$-modules, as in the tensor induction construction. If Y is a free resolution of $V^{\uparrow G}_{\otimes}$ as a $\mathbb{Z}G$-module, we know that there exist chain maps $Y \xrightarrow{\phi} \bigotimes_j(X \otimes g_j)$, and that any two such are homotopic. Thus for i even, if $x \in \mathrm{Hom}_{\mathbb{Z}H}(X_i,W)$, $\bigotimes_j(x \otimes g_j) \in \mathrm{Hom}_{\mathbb{Z}G}(\bigotimes_j(X \otimes g_j), W^{\uparrow G}_{\otimes})$, and composing with ϕ gives

$$\mathrm{Norm}_{H,G}(x) = \phi.\bigotimes_j (x \otimes g_j) \in \mathrm{Hom}_{\mathbb{Z}G}(Y_{|G:H|i}, W^{\uparrow G}_{\otimes}).$$

It is shown in Lemma 4.1.1 of Benson, Representations and Cohomology, II: Cohomology of groups and modules (CUP, 1991) [the original reasoning in these notes was incorrect at this point] that $\mathrm{Norm}_{H,G}$ sends cocycles to cocycles and coboundaries to coboundaries, and hence induces a well defined map for i even

$$\mathrm{norm}_{H,G}:\mathrm{Ext}^i_{kH}(V,W) \to \mathrm{Ext}_{kG}^{|G:H|i}(V^{\uparrow G}_{\otimes}, W^{\uparrow G}_{\otimes}).$$

These maps satisfy the Mackey type formulae

$$\mathrm{res}_{G,K}\mathrm{tr}_{H,G}(x) = \sum_{HgK} \mathrm{tr}_{H^g \cap K,K}(\mathrm{res}_{H^g,H^g \cap K}(x^g))$$

$$\text{(if } V = W = k\text{)}\quad \mathrm{res}_{G,K}\mathrm{norm}_{H,G}(x) = \prod_{HgK} \mathrm{norm}_{H^g \cap K,K}(\mathrm{res}_{H^g,H^g \cap K}(x^g)).$$

(ix) Let $P \in \mathrm{Syl}_p(G)$. Then $\mathrm{tr}_{P,G}\mathrm{res}_{G,P}(x) = |G:P|.x$, and so $\mathrm{res}_{G,P}$ is injective.

(Warning: it is not in general true that $\mathrm{norm}_{H,G}\mathrm{res}_{G,H}(x)=x^{|G:H|}$)

(x) <u>Shapiro's lemma</u>

If V is a $\mathbb{Z}H$-module then there is a natural isomorphism $H^i(G,V{\uparrow}^G) \cong H^i(H,V)$; more generally, if U is a $\mathbb{Z}G$-module and V is a $\mathbb{Z}H$-module then there are natural isomorphisms $\mathrm{Ext}^i_G(U,V{\uparrow}^G) \cong \mathrm{Ext}^i_H(U{\downarrow}_H,V)$ and $\mathrm{Ext}^i_G(V{\uparrow}^G,U) \cong \mathrm{Ext}^i_H(V,U{\downarrow}_H)$. These are proved by induction on i, starting with Frobenius reciprocity and using the long exact sequence in cohomology.

Now suppose $N \trianglelefteq G$, and V is a $\mathbb{Z}G$-module. Our intention is to develop the Lyndon-Hochschild-Serre spectral sequence comparing $H^{p+q}(G,V)$ with $H^p(G/N,H^q(N,V))$. We use the normalized bar resolution, and we let $V^i = \mathrm{Hom}_{\mathbb{Z}G}(\tilde{X}_i(G),V)$ for $i \geq 0$ as above. We filter V^i by

$$F^pV^i = \begin{cases} V^i & p \leq 0 \\ \{f \in V^i : [g_1|..|g_i]f = 0 \text{ whenever at least } i-p+1 \text{ of the } g_j \text{ are in } N\}, & 0 < p \leq i \\ 0 & i < p \end{cases}$$

It is easy to see that $(F^pV^i)\delta^i \subseteq F^pV^{i+1}$ and $F^pV^i \supseteq F^{p+1}V^i$, and that the cup product map takes $F^pU^i \otimes F^qV^j$ into $F^{p+q}(U \otimes V)^{i+j}$.

We also introduce a second filtration of V^i as follows.

$$\tilde{F}^pV^i = \begin{cases} V^i & p \leq 0 \\ \{f \in V^i : [g_1|..|g_i]f \text{ depends only on the cosets } Ng_j \text{ for } j \leq p\}, & 0 < p \leq i \\ 0 & i < p \end{cases}$$

This also satisfies $(\tilde{F}^pV^i)\delta^i \subseteq \tilde{F}^pV^{i+1}$ and $\tilde{F}^pV^i \supseteq \tilde{F}^{p+1}V^i$. This filtration has the disadvantage that it is not compatible with the cup product, but the advantage that it is easier to calculate with. The fact that they give rise to the same spectral sequence will follow from the following lemma.

2.22.2 Lemma

$H^j(F^pV^i/\tilde{F}^pV^i,\delta) = 0.$

Proof

This follows from an explicit calculation with cocycles and coboundaries. Note that $\tilde{F}^pV^i \subseteq F^pV^i$ since any subset of $\{1,\ldots, i\}$ of size at least $i-p+1$ contains an element of $\{1,\ldots, p\}$. □

2.22.3 Proposition

The inclusion $\tilde{F}^pV^i \subseteq F^pV^i$ induces isomorphisms

(i) $\Psi: H^j(F^pV^i,\delta) \cong H^j(\tilde{F}^pV^i,\delta)$

(ii) $\Phi: H^j(F^pV^i/F^{p+1}V^i,\delta) \cong H^j(\tilde{F}^pV^i/\tilde{F}^{p+1}V^i,\delta)$

Proof

(i) follows from 2.22.2 and the long exact sequence of cohomology.

(ii) follows from (i) and the five-lemma. □

We now construct the spectral sequence. Set

$$D_o^{p,q} = D_o^{p,q}(V) = \begin{cases} F^pV^{p+q} & p+q \geq 0 \\ 0 & p+q < 0 \end{cases}$$

$$E_o^{p,q} = E_o^{p,q}(V) = D_o^{p,q}/D_o^{p+1,q-1} \quad .$$

Note that $E_o^{p,q} = 0$ whenever $p < 0$ or $q < 0$. Set

$$\begin{aligned} D_1^{p,q} = D_1^{p,q}(V) &= H^{p+q}(D_o^{p,q}(V),\delta) \\ &= \mathrm{Ker}(\delta^{p+q}|_{D_o^{p,q}})/\mathrm{Im}(\delta^{p+q-1}|_{D_o^{p-1,q+1}}) \\ &= H^{p+q}(F^pV^{p+q},\delta) \\ &\cong H^{p+q}(\tilde{F}^pV^{p+q},\delta) \end{aligned}$$

$$\begin{aligned} E_1^{p,q} = E_1^{p,q}(V) &= H^{p+q}(E_o^{p,q},\delta) \\ &= H^{p+q}(F^pV^{p+q}/F^{p+1}V^{p+q},\delta) \\ &\cong H^{p+q}(\tilde{F}^pV^{p+q}/\tilde{F}^{p+1}V^{p+q},\delta). \end{aligned}$$

Then the short exact sequence

$$0 \to D_o^{p+1,q-1} \to D_o^{p,q} \to E_o^{p,q} \to 0$$

gives rise to a long exact sequence

$$0 \to D_1^{p+1,q-1} \xrightarrow{i_1^{p,q}} D_1^{p,q} \xrightarrow{j_1^{p,q}} E_1^{p,q} \xrightarrow{k_1^{p,q}}$$

$$D_1^{p+1,q} \xrightarrow{i_1^{p+1,q}} D_1^{p,q+1} \xrightarrow{j_1^{p+1,q}} E_1^{p+1,q} \xrightarrow{k_1^{p+1,q}}$$

$$D_1^{p+1,q+1} \longrightarrow \cdots$$

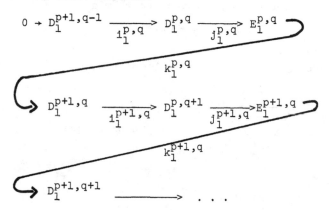

Notice that by this stage it does not matter whether we started with $\tilde{F}^pV^{p+q}$ or F^pV^{p+q} .

Setting $\quad D_1 = D_1(V) = \bigoplus_{p,q} D_1^{p,q}(V)$

and $\quad E_1 = E_1(V) = \bigoplus_{p,q} E_1^{p,q}(V)$

we fit these homomorphisms together to make an <u>exact couple</u>

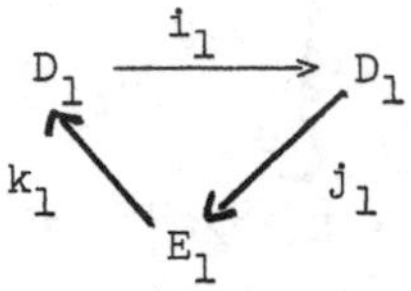

i.e. each pair of consecutive maps is exact.

Every time we have an exact couple as above, we obtain a spectral sequence as follows. The spectral sequence arising from the particular exact couple we have described is called the Lyndon-Hochschild-Serre spectral sequence.

Since $j_1 k_1 = 0$, we have $(k_1 j_1)^2 = 0$. Thus setting $d_1 = k_1 j_1$, (E_1, d_1) is a cochain complex. We define

$$D_2 = \mathrm{Im}(i_1)$$

$$E_2 = H(E_1, d_1)$$

$$i_2 = i_1|_{D_2} : D_2 \to D_2$$

If $x \in D_2$, write $x = yi_1$ and define $xj_2 = yj_1 + \mathrm{Im}(d_1)$ to obtain

$$j_2 : D_2 \to E_2$$

If $z + \mathrm{Im}(d_1) \in E_2$, define $(z + \mathrm{Im}(d_1))k_2 = zk_1$ to obtain

$$k_2 : E_2 \to D_2.$$

2.22.4 Lemma

The maps j_2 and k_2 are well defined, and

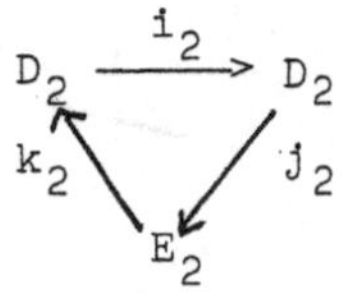

is an exact couple.

Proof

Easy diagram chasing. □

The couple D_2, E_2, i_2, j_2, k_2 is called the derived couple of D_1, E_1, i_1, j_1, k_1. Continuing this way, we obtain exact couples D_n, E_n, i_n, j_n, k_n for each $n \geq 1$. D_n and E_n are bigraded as follows.

$$D_n^{p,q} = \mathrm{Im}(i_{n-1} : D_{n-1}^{p+1,q-1} \to D_{n-1}^{p,q})$$

$$E_n^{p,q} = \frac{\mathrm{Ker}(d_{n-1}: E_{n-1}^{p,q} \to E_{n-1}^{p+n-1,q-n+2})}{\mathrm{Im}(d_{n-1} : E_{n-1}^{p-n+1,q+n-2} \to E_{n-1}^{p,q})}$$

i_n has bidegree $(-1,1)$

j_n has bidegree $(n-1,-n+1)$

k_n has bidegree $(1,0)$

$d_n = k_n j_n$ has bidegree $(n,-n+1)$.

Each $D_n^{p,q}$ is a submodule of $D_o^{p,q}$, and we write $D_\infty^{p,q}$ for $\bigcap_n D_n^{p,q}$. Since each $E_n^{p,q}$ is a subquotient of $E_{n-1}^{p,q}$ we may find subgroups $D_o^{p,q} = Z_o^{p,q} \supseteq Z_1^{p,q} \supseteq Z_2^{p,q} \supseteq \ldots \supseteq B_2^{p,q} \supseteq B_1^{p,q} \supseteq B_o^{p,q} = D_o^{p+1,q-1}$ such that $E_n^{p,q} = Z_n^{p,q}/B_n^{p,q}$. We set $Z_\infty^{p,q} = \bigcap_n Z_n^{p,q}$, $B_\infty^{p,q} = \bigcup_n B_n^{p,q}$ and $E_\infty^{p,q} = Z_\infty^{p,q}/B_\infty^{p,q}$. Note that since $E_n^{p,q} = 0$ whenever $p < 0$ or $q < 0$ and since d_n has bidegree $(n,-n+1)$, it follows that for $n > \max(p,q) + 1$, $E_n^{p,q} = E_{n+1}^{p,q} = E_\infty^{p,q}$. (In fact it can be shown [50] that there exists a value of n independent of p and q such that $E_n^{p,q} = E_{n+1}^{p,q} = E_\infty^{p,q}$).

<u>Remark</u>

The maps $d_n: E_n^{o,n-1} \to E_n^{n,o}$ are called the <u>transgressions</u> or <u>face maps</u>.

2.22.5 <u>Theorem</u>

(i) $E_1^{p,q}(V) \cong \mathrm{Hom}_{G/N}(\tilde{X}_p(G/N), H^q(N,V))$

(ii) $E_2^{p,q}(V) \cong H^p(G/N, H^q(N,V))$

(iii) $H^{p+q}(G,V)$ has a filtration $F^pH^{p+q}(G,V)$ such that

$$F^pH^{p+q}(G,V)/F^{p+1}H^{p+q}(G,V) \cong E_\infty^{p,q}(G,V) .$$

<u>Sketch of Proof</u>

(i) We have a homomorphism

$$\rho^{p,q} : \tilde{F}^pV^{p+q} \to \mathrm{Hom}_{G/N}(\tilde{X}_p(G/N), \mathrm{Hom}_N(\tilde{X}_q(N),V))$$

given as follows. If $\varphi \in \tilde{F}^pV^{p+q}$,

$$[n_1|\ldots|n_q]([\bar{g}_1|\ldots|\bar{g}_p]\ (\varphi\rho^{p,q})) = [g_1|\ldots|g_p|n_1|\ldots|n_q]\ \varphi.$$

$\rho^{p,q}$ induces a map

$$\rho_1^{p,q} : E_1^{p,q}(V) \to \mathrm{Hom}_{G/N}(\tilde{X}_p(G/N), H^q(N,V))$$

and a somewhat lengthy calculation shows that $\rho_1^{p,q}$ is an isomorphism.

(ii) A similar calculation shows that

$$d_1^{p,q}\ \rho_1^{p,q} = (-1)^q\ \rho_1^{p-1,q}\ \delta$$

where δ is the coboundary homomorphism for the complex $\mathrm{Hom}_{G/N}(\tilde{X}_p(G/N), H^q(N,V))$. Thus $\rho_1^{p,q}$ induces an isomorphism from $E_2^{p,q}(V) = H(E_1^{p,q}(V), d_1)$ to $H^p(G/N, H^q(N,V))$ $= H(\mathrm{Hom}_{G/N}(\tilde{X}_p(G/N), H^q(N,V)), \delta)$.

(iii) For $n > \max(p,q) + 1$, we have $E_n^{p,q} = E_\infty^{p,q}$, and the exact sequence

$$\cdots \xrightarrow[k_n]{} D_n^{p-n+2,q+n-2} \xrightarrow[i_n]{} D_n^{p-n+1,q+n-1} \xrightarrow[j_n]{} E_n^{p,q} \xrightarrow[k_n]{} D_n^{p+1,q} \longrightarrow \cdots$$

reduces to

$$0 \to \mathrm{Im}(H^{p+q}(\tilde{F}^{p+1}V^{p+q}) \to H^{p+q}(G,V)) \to \mathrm{Im}(H^{p+q}(\tilde{F}^pV^{p+q}) \to H^{p+q}(G,V)) \to E_\infty^{p,q} \to 0$$

Thus if we filter $H^{p+q}(G,V)$ by letting

$$F^pH^{p+q}(G,V) = \mathrm{Im}(H^{p+q}(\tilde{F}^pV^{p+q}) \to H^{p+q}(G,V))$$

we have

$$F^oH^{p+q}(G,V) = H^{p+q}(G,V)$$

$$F^pH^{p+q}(G,V)\ /F^{p+1}H^{p+q}(G,V) \cong E_\infty^{p,q}(V)$$

$$F^{p+q+1}H^{p+q}(G,V) = 0. \quad \square$$

We express the information given in the above theorem by writing

$$H^p(G/N, H^q(N,V)) \Rightarrow H^{p+q}(G,V).$$

Another computation similar in nature to all the others we have avoided writing out shows that the natural cup-product structures on the two sides of 2.22.5(i) are related by

$$(u \cup v)\ \rho_1^{p+p',q+q'} = (-1)^{p'q}(u\ \rho_1^{p,q} \cup v\ \rho_1^{p',q'}),$$

$(u \in E_1^{p,q}(V),\ v \in E_1^{p,q}(V),\ u \cup v \in E_1^{p,q}(U \otimes V))$.

Since the cup product on E_1 satisfies

$$(u \cup v)d_1 = u \cup vd_1 + (-1)^{\deg(v)}\ ud_1 \cup v$$

(where $\deg(v) = p' + q'$ if $v \in E_1^{p',q'}$), it follows that a cup product structure on E_2 is induced, and so on, at each stage satisfying

$$(u \cup v)d_r = u \cup vd_r + (-1)^{\deg(v)}\, ud_r \cup v.$$

The cup product structure at the E_∞ level is just the graded version of the cup product $H^*(G,U) \otimes H^*(G,V) \to H^*(G,U \otimes V)$.

Setting V equal to the trivial module k, we get maps

$$E_r^{p,q}(U) \otimes E_r^{p',q'}(k) \to E_r^{p+p',q+q'}(U)$$

making $E_r(U)$ into a module over $E_r(k)$, and likewise $E_\infty(U)$ into a module over $E_\infty(k)$.

As an example of an easy application of the spectral sequence we give the following.

2.22.6 <u>Proposition</u>

There is a five term exact sequence

$$0 \to H^1(G/N,V^N) \to H^1(G,V) \to H^1(N,V)^{G/N} \to H^2(G/N,V^N) \to H^2(G,V)$$

<u>Proof</u>

By 2.22.5 we are looking for maps

$$0 \to E_2^{1,0} \to \begin{array}{|c|}\hline E_\infty^{0,1} \\ \hline E_\infty^{1,0} \\ \hline\end{array} \to E_2^{0,1} \to E_2^{2,0} \to \begin{array}{|c|}\hline E_\infty^{0,2} \\ \hline E_\infty^{1,1} \\ \hline E_\infty^{2,0} \\ \hline\end{array}$$

We have at the E_2 level the following maps

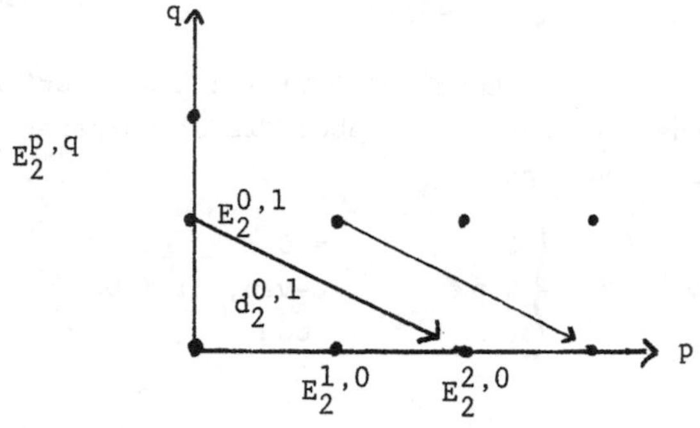

Thus

$$E_2^{1,0} \cong E_\infty^{1,0}$$
$$E_\infty^{0,1} \cong \mathrm{Ker}(d_2^{0,1})$$
$$E_\infty^{2,0} \cong \mathrm{Coker}(d_2^{0,1})$$

and the proposition is proved. □

2.22.7 Corollary

Suppose G is a p-group, and the map $G \to G/\Phi(G)$ induces a monomorphism $H^2(G/\Phi(G),\mathbb{Z}/p\mathbb{Z}) \to H^2(G,\mathbb{Z}/p\mathbb{Z})$. Then $\Phi(G) = 1$, i.e. G is elementary abelian.

Proof

Since $H^1(G/\Phi(G),\mathbb{Z}/p\mathbb{Z}) \cong H^1(G,\mathbb{Z}/p\mathbb{Z})$ ($\cong [G/\Phi(G)]^*$, see exercise 1) the above exact sequence implies that $H^1(\Phi(G),\mathbb{Z}/p\mathbb{Z})^{G/\Phi(G)} = 0$. Hence $\Phi(G)/\Phi^2(G) = 1$, and so $\Phi(G) = 1$. □

Exercises

1. Let G be a p-group. Using the isomorphism $H^1(G,\mathbb{Z}/p\mathbb{Z}) \cong \mathrm{Ext}^1_G(\mathbb{Z}/p\mathbb{Z},\mathbb{Z}/p\mathbb{Z})$ show that $H^1(G,\mathbb{Z}/p\mathbb{Z}) \cong [G/\Phi(G)]^* \cong (J((\mathbb{Z}/p\mathbb{Z})G)/J^2((\mathbb{Z}/p\mathbb{Z})G))^*$ (hint: look at matrices of shape $\begin{pmatrix}1 & * \\ 0 & 1\end{pmatrix}$ over $\mathbb{Z}/p\mathbb{Z}$).

2. Let G be a cyclic group of order p, $G = \langle g\rangle$, and let V be a $\mathbb{Z}G$-module. Using the resolution

$$0 \longleftarrow \mathbb{Z} \xleftarrow{\ \varepsilon\ } (\mathbb{Z}G)_0 \xleftarrow{\ \partial_1\ } (\mathbb{Z}G)_1 \xleftarrow{\ \partial_2\ } (\mathbb{Z}G)_2 \xleftarrow{\ \partial_3\ } \cdots$$

given by

$$\varepsilon : (1)_0 \to 1$$
$$\partial_i : (1)_i \to \begin{cases}(1 + g + \cdots + g^{p-1})_{i-1} & i \text{ even} \\ (1 - g)_{i-1} & i \text{ odd}\end{cases}$$

calculate $H^i(G,V)$ (it is clear that $H^i(G,V)$ is a subquotient of V since each term in the free resolution above is a one-generator module). In particular, show that

$$H^i(G,\mathbb{Z}) \cong \begin{cases}\mathbb{Z} & i = 0 \\ \mathbb{Z}/p\mathbb{Z} & i \text{ even, } i \neq 0 \\ 0 & i \text{ odd}\end{cases}$$

and

$$H^i(G,\mathbb{Z}/p\mathbb{Z}) \cong \mathbb{Z}/p\mathbb{Z} \quad \text{for all } i$$

(compare this with the universal coefficient theorem).
Write down the long exact sequence associated with the short exact sequence $0 \to \mathbb{Z} \to \mathbb{Z} \to \mathbb{Z}/p\mathbb{Z} \to 0$ of coefficients.

Using the diagonal approximation

$$(1)_{q+r}\Delta = \Sigma (1)_{q+r}\Delta_{q,r}$$

$$(1)_{q+r}\Delta_{q,r} = \begin{cases} (1)_q \otimes (1)_r & r \text{ even} \\ (g)_q \otimes (1)_r & q \text{ even, } r \text{ odd} \\ -\sum_{0\le i<j\le p-1} (g^j)_q \otimes (g^i)_r & q \text{ odd, } r \text{ odd} \end{cases}$$

(check that this is indeed a diagonal approximation; you will need the identity

$$(1 \otimes 1 - g \otimes g)(\sum_{0\le i<j\le p-1} g^j \otimes g^i)$$

$$= (1+g+..+g^{p-1}) \otimes 1 - 1 \otimes (1+g+..+g^{p-1}))$$

show that the ring structure of $H^*(G,\mathbb{Z}/p\mathbb{Z})$ is as follows.

(i) $p \neq 2$:

generators u, v

$\deg(u) = 1$, $\deg(v) = 2$

$u^2 = 0$, $uv = vu$ (i.e. $E(u) \otimes P(v)$)

(ii) $p = 2$:

generator v

$\deg(v) = 1$

no relations (i.e. $P(v)$).

Show that if α is a generator for $H^2(G,\mathbb{Z}/p\mathbb{Z})$ then for any G-module V, multiplication by α yields an epimorphism $H^o(G,V) \to H^2(G,V)$ and isomorphisms $H^q(G,V) \to H^{q+2}(G,V)$ for $q > 0$.

3. Use the Künneth formula to calculate the cohomology ring of an arbitrary elementary abelian group.

4. Let $P \in \mathrm{Syl}_p(G)$, and let V be a kG-module. Suppose P is a t.i. subgroup of G (i.e. for $g \in G$, either $P^g = P$ or $P \cap P^g = 1$), with normalizer N. Show that $\mathrm{res}_{G,N}$ is an isomorphism between $H^*(G,V)$ and $H^*(N,V)$. (Hint: use $\mathrm{tr}_{N,G}$ and the Mackey formula).

5. Calculate $H^*(A_5,k)$ for k an algebraically closed field of

(i) characteristic 2

(ii) characteristic 3

(iii) characteristic 5.

(Hint: use question 2 to calculate $H^*(P,k)$ for $P \in \mathrm{Syl}_p(A_5)$, then use a spectral sequence to calculate $H^*(N_G(P),k)$, and finally use question 4 to complete the calculation).

6. (P. Webb)

With the constants λ_H as in 2.13.6, show that if U and V are modules for G then

(i) $$\dim_k \mathrm{Ext}^n_G(U,V) = \sum_{H \in \mathrm{Hyp}_p(G)} \lambda_H \dim_k \mathrm{Ext}^n_H(U{\downarrow}_H, V{\downarrow}_H)$$

(ii) $$\dim_k H^n(G,V) = \sum_{H \in \mathrm{Hyp}_p(G)} \lambda_H \dim_k H^n(H,V)$$

(iii) $$\xi_V(t) = \sum_{H \in \mathrm{Hyp}_p(G)} \lambda_H \; \xi_{V{\downarrow}_H}(t)$$

(in these sums, H runs over a set of representatives of conjugacy classes of p-hypoelementary subgroups).

7. Let G be a p-group and V a kG-module. Using the long exact sequence of cohomology, show that

(i) $|G|^{-1}\dim_k \Omega(V) \le \dim_k V \le |G| \dim_k \Omega(V)$

(ii) $|G|^{-2}\dim_k (V) \le |G|^{-1}\dim_k \mho(V) \le \dim_k H^1(G,V)$
$\le \dim_k \Omega(V) \le |G| \dim_k V$

(iii) Given $n > 0$,
$$|G|^{-n-1}\dim_k V \le \dim_k H^n(G,V) \le |G|^n \dim_k V$$

(use $H^n(G,V) \cong H^1(G, \mho^{n-1}(V))$)

(iv) Given $m, n > 0$,
$$\dim_k H^m(G,V) \le |G|^{|m-n|+3} \dim_k H^n(G,V).$$

2.23 Bockstein Operations and the Steenrod Algebra

In this section we describe the operations in cohomology necessary for the study of complexity theory in section 2.24. We begin with the Bockstein operations.

The short exact sequence
$$0 \to \mathbb{Z} \overset{\lambda}{\to} \mathbb{Z} \overset{\mu}{\to} \mathbb{Z}/p\mathbb{Z} \to 0,$$

where the left-hand map is given by multiplication by p, gives a long exact sequence in cohomology

$$\cdots \xrightarrow{\lambda_q} H^q(G,\mathbb{Z}) \xrightarrow{\mu_q} H^q(G,\mathbb{Z}/p\mathbb{Z}) \xrightarrow{\nu_q} H^{q+1}(G,\mathbb{Z}) \xrightarrow{\lambda_{q+1}} \cdots .$$

We define $\beta_q = \nu_q\mu_{q+1} : H^q(G,\mathbb{Z}/p\mathbb{Z}) \to H^{q+1}(G,\mathbb{Z}/p\mathbb{Z})$. This map β is called the Bockstein map. It is easy to verify that the following are satisfied.

(i) $\beta^2 = 0$

(ii) $(xy)\beta = x(y\beta) + (-1)^{\deg(y)}(x\beta)y$.

We shall also need the Steenrod operations on cohomology. These operations are in fact quite difficult to construct, and so we shall be content to list their properties and to take their existence for granted.

2.23.1 Theorem

There exist unique operations $P^i : H^n(G,\mathbb{Z}/p\mathbb{Z}) \to H^{n+2i(p-1)}(G,\mathbb{Z}/p\mathbb{Z})$ (called the Steenrod operations, or reduced power operations) satisfying axioms (i) - (v):

(i) P^i is a natural transformation of functors.

(ii) $P^o = 1$

In case $p = 2$ we write Sq^{2i} for P^i and Sq^{2i+1} for $P^i\beta$. The Sq^i are called the Steenrod Squares.

(iii) $(p \neq 2)$ If $\deg(x) = 2n$ then $xP^n = x^p$

$(p = 2)$ If $\deg(x) = n$ then $xSq^n = x^2$

(iv) $(p \neq 2)$ If $\deg(x) < 2n$ then $xP^n = 0$

$(p = 2)$ If $\deg(x) < n$ then $xSq^n = 0$

(v) Cartan formula If $p \neq 2$, $(xy)P^n = \sum_{i=0}^{n} (xP^i)(yP^{n-i})$

If $p = 2$, $(xy)Sq^n = \sum_{i=0}^{n} (xSq^i)(ySq^{n-i})$.

The axioms (i) - (v) imply

(vi) Adem Relations, $p \neq 2$

If $b < pa$ then

$$P^aP^b = \sum_{j=0}^{[b/p]} (-1)^{b+j} \binom{(p-1)(a-j)-1}{b-pj} P^jP^{a+b-j}$$

If $b \leq a$ then

$$P^a\beta P^b = \sum_{j=0}^{[b/p]} (-1)^{b+j} \binom{(p-1)(a-j)}{b-pj} P^j P^{a+b-j}\beta$$

$$+ \sum_{j=0}^{[(b-1)/p]} (-1)^{b+j-1} \binom{(p-1)(a-j)-1}{b-pj-1} P^j \beta P^{a+b-j}$$

(vii) Adem relations , $p = 2$

If $0 < b < 2a$ then

$$Sq^a Sq^b = \sum_{j=0}^{[b/2]} \binom{a-1-j}{b-2j} Sq^j Sq^{a+b-j}$$

(viii) For $p \neq 2$, $x \in H^1(G,\mathbb{Z}/p\mathbb{Z}) \Rightarrow xP^i = 0$ for $i > 0$.

(ix) For $p \neq 2$, $x \in H^2(G,\mathbb{Z}/p\mathbb{Z}) \Rightarrow x^j P^i = \binom{j}{i} x^{j+(p-1)i}$

Proof

See [89] and [90]. See also exercises 1 and 2 for a sketch of the construction, and [99] for a more extensive discussion. □

The algebra generated by the P^i and β subject to the Adem relations is called the Steenrod Algebra $\mathbb{A}(p)$. Thus $\mathbb{A}(p)$ has a natural action on $H^*(G,\mathbb{Z}/p\mathbb{Z})$. See [68] for a discussion of $\mathbb{A}(p)$ and its dual.

Let $T = \sum_{i=0}^{\infty} P^i$ if $p \neq 2$, and $T = \sum_{i=0}^{\infty} Sq^i$ if $p = 2$. By axiom (iv), T has a well defined action on $H^*(G,\mathbb{Z}/p\mathbb{Z})$ since for given x, only finitely many of the xP^i are non-zero. Moreover, the Cartan formula shows that T is an algebra homomorphism.

2.23.2 Lemma

Let $x \in H^2(G,\mathbb{Z}/p\mathbb{Z})$. Then

$$xT = \begin{cases} x + x^p & p \neq 2 \\ x + x^p + x\beta & p = 2 \end{cases}$$

Proof

For $p \neq 2$, this follows from 2.23.1(ix). For $p = 2$, it follows from (iii) and (iv). □

We are now ready to prove a theorem of Serre on Bocksteins for p-groups.

2.23.3 Theorem (Serre)

Suppose G is a p-group. If G is not elementary abelian, then there are elements $x_1, \ldots, x_r$ of $[G/\Phi(G)]^* \cong H^1(G,\mathbb{Z}/p\mathbb{Z})$ such that $(x_1\beta) \ldots (x_r\beta) = 0$ as an element of $H^{2r}(G,\mathbb{Z}/p\mathbb{Z})$.

<u>Proof</u>

If G is not elementary abelian, then 2.22.7 tells us that the map $H^2(G/\Phi(G), \mathbb{Z}/p\mathbb{Z}) \to H^2(G, \mathbb{Z}/p\mathbb{Z})$ is not injective. By exercise 2 of 2.22 and the Künneth formula, if $y_1, \ldots, y_n$ form a basis for $H^1(G/\Phi(G), \mathbb{Z}/p\mathbb{Z})$ then $\{y_i\beta\} \cup \{y_i y_j, i < j\}$ form a basis for $H^2(G/\Phi(G), \mathbb{Z}/p\mathbb{Z})$. Thus there is a non-trivial linear relation

$$(*) \qquad \sum_{i<j} a_{ij} y_i y_j + \sum_k b_k(y_k\beta) = 0$$

as elements of $H^2(G, \mathbb{Z}/p\mathbb{Z})$. If all the a_{ij} are zero, take $r = 1$ and $x_1 = \Sigma b_k y_k$. Thus we may assume that some a_{ij} is non-zero. Applying the element $\beta P^1 \beta \in A(p)$ to the relation (*), and using the relations 2.23.1 (v), (viii), (ix) and the relation $\beta^2 = 0$, we obtain

$$\sum_{i<j} a_{ij}((y_i\beta)^p(y_j\beta) - (y_i\beta)(y_j\beta)^p) = 0.$$

Note that by 2.22.1(iii), the $y_i\beta$ commute.

Let $\hat{k}$ be an algebraic closure of $\mathbb{Z}/p\mathbb{Z}$, and denote by $\mathfrak{U}$ the (homogeneous) ideal of $\hat{k}[X_1, \ldots, X_n]$ generated by the relations among $y_1\beta, \ldots, y_n\beta$. Thus we have shown that $\mathfrak{U} \neq 0$. Moreover, by 2.23.2, $\mathfrak{U}$ is stable under the operation of replacing X_i by $X_i + X_i^p$ (since $(y_i\beta)\beta = 0$). Denote by $\mathfrak{V}$ the variety in $\hat{k}^n$ defined by $\mathfrak{U}$, and denote by F the Frobenius map $(\lambda_1, \ldots, \lambda_n)F = (\lambda_1^p, \ldots, \lambda_n^p)$ on $\hat{k}^n$. Then if $v \in \mathfrak{V}$, so is $\lambda(v + vF)$ for any constant λ.

We show by induction on i that the linear subspace $W_i(v)$ spanned by $v, vF, \ldots, vF^i$ is in $\mathfrak{V}$. Suppose true for $i-1$.

Let $0 \neq w = w_1 v + w_2(vF) + \cdots + w_i(vF^i) \in W_i(v)$. Let λ be a solution of the algebraic equation

$$\lambda^{p^{i-1}}\left(\frac{w_i}{\lambda} - \left(\frac{w_{i-1}}{\lambda}\right)^p + \cdots \pm \left(\frac{w_1}{\lambda}\right)^{p^{i-1}}\right) = 0,$$

let $u_j = \left(\frac{w_j}{\lambda}\right) - \left(\frac{w_{j-1}}{\lambda}\right)^p + \cdots \pm \left(\frac{w_1}{\lambda}\right)^{p^{j-1}}$ for $1 \leq j \leq i-1$, and let $u = \sum_{j=1}^{i-1} u_j(vF^j) \in W_{i-1}(v)$. Then it is easy to check that

$w = \lambda(u + uF) \in \mathfrak{V}$. Hence $W_i(v) \subseteq \mathfrak{V}$ and so $W(v) = \bigcup_i W_i(v) \subseteq \mathfrak{V}$.

This means that $\mathfrak{V}$ is a union of subspaces stable under F, and hence $\mathfrak{V}$ is contained in the union of all hyperplanes stable under F, of which there are only finitely many. These hyperplanes are defined by the equations represented by $x_i\beta = 0$ for the non-zero elements $x_i \in H^1(G,\mathbb{Z}/p\mathbb{Z})$. Thus $\prod_i(x_i\beta)$ represents an equation which vanishes on $\mathfrak{v}$. Thus by Hilbert's Nullstellensatz, some power of $\prod_i(x_i\beta)$ is in $\mathfrak{u}$, thus proving the theorem. □

Finally, we examine the relationship between the Bockstein operations and the Lynson-Hochschild-Serre spectral sequence.

2.23.4 Proposition (Quillen, Venkov)

Let G be a p-group and V an kG-module. Let $x\beta$ be the Bockstein of an element x of $H^1(G,\mathbb{Z}/p\mathbb{Z}) \subseteq H^1(G,k)$. Regarding x as an element of $[G/\Phi(G)]^*$, let H be the corresponding maximal subgroup of G. Then

$$F^pH^{p+q}(G,V).(x\beta) = F^{p+2}H^{p+q+2}(G,V)$$

in the filtration arising from the spectral sequence

$$H^p(G/H,H^q(H,V)) \Rightarrow H^{p+q}(G,V).$$

Proof

Let $\bar{x} \in H^1(G/H,k)$ be the element corresponding to x. Then $\bar{x}\beta \in E_2^{2,0}(k)$, and since $d_i^{2,0} = 0$ for all i, $\bar{x}\beta$ has images y_i in $E_i^{2,0}(k)$ for all i. If $v \in E_i^{p,q}(V)$ then by 2.22.1(i),

$$(vy_i)d_i = v(y_id_i) + (-1)^2(vd_i)y_i = (vd_i)y_i$$

Thus the map

$$b_i^{p,q} : E_i^{p,q}(V) \to E_i^{p+2,q}(V)$$

given by multiplication by y_i commutes with d_i.

We shall prove by induction on i that

(a) $b_i^{p,q}$ is an epimorphism for $p \geq 0$

(b) $b_i^{p,q}$ is an isomorphism for $p \geq i-1$ (see diagram) .

Case 1 $i = 2$

In this case we are looking at

$$H^p(G/H,H^q(H,V)) \to H^{p+2}(G/H,H^q(H,V))$$

and since G/H is cyclic of order p, the result follows from exercise 2 of 2.22.

Case 2 $i > 2$, and (a) and (b) are true up to $i-1$.

(a) Given $\bar{v} \in E_i^{p+2,q}$, choose an inverse image v in $E_{i-1}^{p+2,q}$, so that $vd_{i-1} = 0$. By the inductive hypothesis (a), we may write $v = ub_{i-1}$, $u \in E_{i-1}^{p,q}$. Then $ud_{i-1}b_{i-1} = ub_{i-1}d_{i-1} = 0$ and

$ud_{i-1} \in E_{i-1}^{p+i-1,q-i+2}$ and so by inductive hypothesis (b), $ud_{i-1} = 0$. Thus $\bar{u} \in E_i^{p,q}$ is an element with $\bar{u}b_i = \bar{v}$.

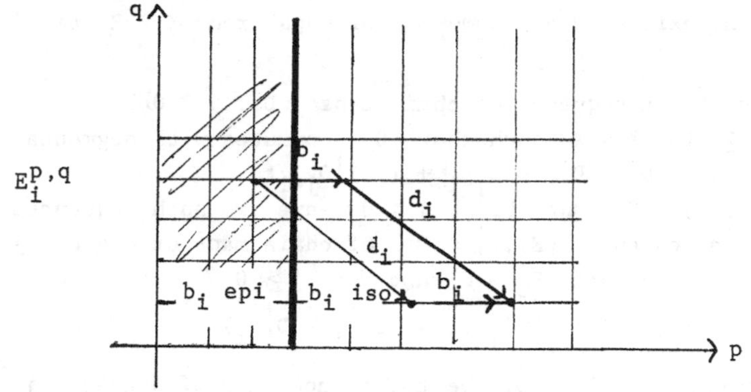

(b) Given $\bar{u} \in E_i^{p,q}$ with $p \geq i-1$, choose an inverse image u in $E_{i-1}^{p,q}$ wuth $ud_{i-1} = 0$. Suppose $\bar{u}b_i = 0$. Then $ub_{i-1} = yd_{i-1}$ for some $y \in E_{i-1}^{p-i+3,q+i}$. Now $p-i+3 \geq 2$, and so we may write $y = zb_{i-1}$, $z \in E_{i-1}^{p-i+1,q+i}$ by the inductive hypothesis (a). Then

$$(u - zd_{i-1})b_{i-1} = yd_{i-1} - zb_{i-1}d_{i-1} = 0$$

and so by the inductive hypothesis (b), $u = zd_{i-1}$, and so $\bar{u} = 0$ in $E_i^{p,q}$.

Having proved (a) and (b), it follows that

$$b^{p,q}: F^pH^{p+q}(G,V)/F^{p+1}H^{p+q}(G,V) \to F^{p+2}H^{p+q+2}(G,V)/F^{p+3}H^{p+q+2}(G,V)$$

given by multiplication by $x\beta$ is an epimorphism. The result follows immediately. □

Exercises

1. (Construction of Sq^i)

Let $T = \langle t : t^2 = 1 \rangle$ be a cyclic group of order 2 and let $\underline{R}$ be the resolution

$$0 \longleftarrow \mathbb{Z} \longleftarrow \mathbb{Z}T \xleftarrow{1-t} \mathbb{Z}T \xleftarrow{1+t} \mathbb{Z}T \xleftarrow{1-t} \cdots$$

of $\mathbb{Z}$ as a $\mathbb{Z}T$-module. Let $\underline{X}$ be a resolution

$$0 \longleftarrow \mathbb{Z} \xleftarrow{\varepsilon} X_0 \xleftarrow{\partial_1} X_1 \xleftarrow{\partial_2} X_2 \xleftarrow{\partial_3} \cdots$$

of $\mathbb{Z}$ as a $\mathbb{Z}G$-module. Then $\underline{X} \otimes \underline{X}$ (defined via the formula (*) in section 2.22) is a (not necessarily free) $\mathbb{Z}(G \times T)$-resolution of $\mathbb{Z}$, where t acts via $(x_1 \otimes x_2)t = (-1)^{\deg(x_1)\deg(x_2)}(x_2 \otimes x_1)$. $\underline{X} \otimes \underline{R}$ is a free $\mathbb{Z}(G \times T)$-resolution of $\mathbb{Z}$, by letting G act on $\underline{X}$ and act on $\underline{R}$. Use the existence and homotopy of maps from $\underline{X} \otimes \underline{R}$ to $\underline{X} \otimes \underline{X}$ to show that

(i) There exists a sequence of chain maps $\{D_j,\ j \geq 0\}$ of degree j from $\underline{X}$ to $\underline{X} \otimes \underline{X}$ such that D_o commutes with augmentation and for $j > 0$, $\partial D_j + D_j \partial = D_{j-1} + (-1)^j D_{j-1} t$.

(ii) If $\{D_j,\ j \geq 0\}$ and $\{D_j',\ j \geq 0\}$ are two such sequences then there exists a sequence $\{E_j,\ j \geq 0\}$ of chain maps of degree j from $\underline{X}$ to $\underline{X} \otimes \underline{X}$ such that $E_o = 0$ and for $j > 0$,
$\partial E_j + (-1)^{j-1} E_j \partial = -E_{j-1} + (-1)^j E_{j-1} t + (D_{j-1} - D_{j-1}')$.

Given a $\mathbb{Z}G$-module V, let D_j^* be the induced map of degree $-j$ from $\mathrm{Hom}_{\mathbb{Z}G}(\underline{X} \otimes \underline{X}, V)$ to $\mathrm{Hom}_{\mathbb{Z}G}(\underline{X}, V)$. If $u \in \mathrm{Hom}_{\mathbb{Z}G}(X_m, V)$ and $v \in \mathrm{Hom}_{\mathbb{Z}G}(X_n, V)$ define $u \smile_i v = (u \otimes v) D_i^* \in \mathrm{Hom}_{\mathbb{Z}G}(X_{m+n-i}, V)$. Show that

$$(u \smile_i v)\delta = (-1)^i u \smile_i v\delta + (-1)^{i+n} u\delta \smile_i v - (-1)^i u \smile_{i-1} v - (-1)^{mn} v \smile_{i-1} u .$$

If $x \in \mathrm{Hom}_{\mathbb{Z}G}(X_n, \mathbb{Z}/2\mathbb{Z})$ define

$$xSq^i = \begin{cases} x \smile_{n-i} x = (x \otimes x) D_{n-i}^* & \text{if } 0 \leq i \leq n \\ 0 & \text{if } i > n. \end{cases}$$

Show that

(i) If $x\delta = 0$ then $(xSq^i)\delta = 0$

(ii) $(x\delta)Sq^i = ((x \otimes x\delta)D_{n-i}^* + (x\delta \otimes x\delta)D_{n-i-1}^*)\ \delta$

(iii) $(x_1 + x_2)Sq^i = x_1 Sq^i + x_2 Sq^i + (x_1 \otimes x_2) D_{n-i+1}^*\ \delta$

(iv) If Sq'^i is defined using another sequence $\{D_j',\ j \geq 0\}$ then $xSq^i - xSq'^i = (x \otimes x) E_{n-i-1}^*\ \delta$.

Deduce that there are well defined operations

$$Sq^i : H^n(G, \mathbb{Z}/2\mathbb{Z}) \to H^{n+i}(G, \mathbb{Z}/2\mathbb{Z}) .$$

(a) If $\deg(x) = n$ then $xSq^n = x^2$ since D_o is a diagonal approximation.

(b) If $\deg(x) > n$ then $xSq^n = 0$ by definition.

(c) If $x \in H^n(G, \mathbb{Z}/2\mathbb{Z})$, choose a cochain $u \in \mathrm{Hom}_{\mathbb{Z}G}(X_n, \mathbb{Z}/4\mathbb{Z})$

whose reduction mod 2 represents x. Then $u\delta$ is in $\mathrm{Hom}_{\mathbb{Z}G}(X_{n+1}, 2\mathbb{Z}/4\mathbb{Z})$ ($\cong \mathrm{Hom}_{\mathbb{Z}G}(X_{n+1}, \mathbb{Z}/2\mathbb{Z})$) and represents $2.x\beta$ in $H^{n+1}(G, \mathbb{Z}/2\mathbb{Z})$. By examining $(u \cup_{n-2j-1} u)\delta$ show that $xSq^{2j}\beta$ is represented by the cocycle

$$(x\beta) \cup_{n-2j-1} u + u \cup_{n-2j-1} (x\beta) + u \cup_{n-2j-2} u$$

Hence $Sq^{2j}\beta = Sq^{2j+1}$ and $Sq^{2j+1}\beta = 0$.

(d) Construct a particular diagonal approximation for the resolution $0 \leftarrow \mathbb{Z}/2\mathbb{Z} \leftarrow (\mathbb{Z}/2\mathbb{Z})T \leftarrow (\mathbb{Z}/2\mathbb{Z})T \leftarrow \ldots$ and use it to show that $(xy)Sq^n = \sum_{i=0}^{n} (xSq^i)(ySq^{n-i})$.

(e) Using the bar resolution, show that the D_j may be chosen so that $[x_1|\ldots|x_j]D_j = [x_1|\ldots|x_j] \otimes [x_1|\ldots|x_j]$. Deduce that $Sq^o = 1$.

<u>Remark</u>

Using the bar resolution of $\mathbb{Z}$ as a $\mathbb{Z}G$-module, we can give explicit maps D_j as follows.

If m is even, we set

$$[x_1|\ldots|x_n]D_m = \sum_{0 \le i_o < i_1 < \ldots < i_m \le n} [x_1|\ldots|x_{i_o}\Big|x_{i_o+1}\ldots x_{i_1}\Big|x_{i_1+1}|\ldots|x_{i_2}\Big|$$
$$x_{i_2+1}\ldots x_{i_3}\Big| \ldots \Big|x_{i_{m-1}+1}|\ldots|x_{i_m}]\, x_{i_m+1}\ldots x_n \otimes [x_{i_o+1}|\ldots|x_{i_1}\Big|$$
$$x_{i_1+1}\ldots x_{i_2}\Big|x_{i_2+1}|\ldots|x_{i_3}\Big| \ldots \Big|x_{i_m+1}|\ldots|x_n] .$$

If m is odd, we set

$$[x_1|\ldots|x_n]\, D_m = \sum_{0 \le i_o < i_1 < \ldots < i_m \le n} [x_{i_o+1}|\ldots|x_{i_1}\Big|x_{i_1+1}\ldots x_{i_2}\Big|$$
$$x_{i_2+1}|\ldots|x_{i_3}\Big| \ldots \Big|x_{i_{m-1}+1}|\ldots|x_{i_m}]x_{i_m+1}\ldots x_n \otimes [x_1|\ldots|x_{i_o}\Big|x_{i_o+1}\ldots x_{i_1}\Big|$$
$$\ldots \Big|x_{i_m+1}|\ldots|x_n]$$

(and zero if $n < m$). Note that D_o is just the Alexander-Whitney map.

2. (Hard) Mimic the above construction for p odd. Let $T = \langle t : t^p = 1\rangle$ and use the existence and homotopy of maps from $\underline{X} \otimes \underline{R}$ to $\otimes^p(\underline{X})$ to construct maps P^i with suitable properties.

Beware that you will also construct the zero operation many times as well!

Remark

Serre ([84] , p. 457) has shown that the Steenrod operations commute with the transgressions in the Lyndon-Hochschild-Serre spectral sequence; you will need to use this fact in the following exercises. For a more general account of how the Steenrod operations fit into spectral sequences, see [82], [86] and [87]. What happens is that Steenrod operations are defined on each page $E_r^{p,q}$ of the spectral sequence. They go up the page (increase q) until property (iv) on the first component of degree tells us that they should be generically zero. Thereafter they go to the right (increase p) with a certain 'indeterminacy' which is killed on a later page of the spectral sequence (there is no indeterminacy at the E_2 and E_∞ levels), until property (iv) on total degree tells us that they should be zero.

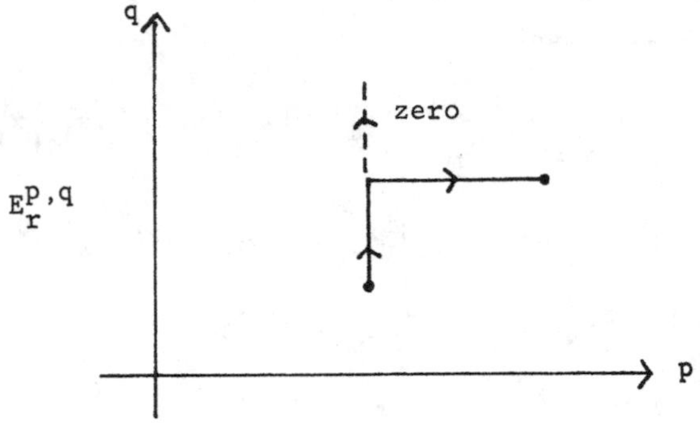

These maps commute with the differentials d_r, and agree with the Steenrod operations we know and love, on the $E_2^{o,*}$, $E_2^{*,o}$ and E_∞ levels. They also satisfy Adem relations and Cartan formulas (v. loc. cit.).

3. Let G be the dihedral group of order eight and let $Z = Z(G)$. Use the spectral sequence $H^p(G/Z, H^q(Z, \mathbb{Z}/2\mathbb{Z})) \Rightarrow H^{p+q}(G, \mathbb{Z}/2\mathbb{Z})$ and the above remark to calculate the ring structure of $H^*(G, \mathbb{Z}/2\mathbb{Z})$.

4. Repeat the above exercise with G equal to the quaternion group of order eight.

Remark

Quillen [74] has shown the following by similar methods. Suppose $0 \to \mathbb{Z}/2\mathbb{Z} \to G \to E \to 0$ is an extension of an elementary abelian two group E by a central subgroup of order two. Since $H^*(E,\mathbb{Z}/2\mathbb{Z}) \cong S(E^*)$, the symmetric algebra on the dual of E as a vector space, the given extension corresponds to a certain quadratic form $Q(x) \in S^2(E^*)$. Let B be the associated bilinear form. Then

$$Q(x).Sq^1Sq^2 \,..\, Sq^{2^{i-1}} = B(x,x^{2^i})$$

for each i. Each $B(x,x^{2^i})$ is a non-zero-divisor modulo the ideal generated by the previous ones in $S(E^*)$, for $1 \le i < h$, h being the codimension of a maximal isotropic subspace of E, and for $i \ge h$ the $B(x,x^{2^i})$ are in the ideal generated by the previous ones. The spectral sequence $H^p(E,H^q(\mathbb{Z}/2\mathbb{Z},\mathbb{Z}/2\mathbb{Z})) \Rightarrow H^{p+q}(G,\mathbb{Z}/2\mathbb{Z})$ therefore converges at the E_{h+1} level, and in fact

$$H^*(G,\mathbb{Z}/2\mathbb{Z}) \cong (S(E^*)/\langle Q(x),B(x,x^2),\, ..\, ,B(x,x^{2^{h-1}})\rangle) \otimes \mathbb{Z}/2\mathbb{Z}[w_{2^h}]$$

where w_{2^h} is an element of degree 2^h (appearing on the left-hand wall of the spectral sequence). See also [100] for a partial analysis of the corresponding case in odd characteristic.

2.24 Complexity

In this section, we shall define the complexity of a module, and develop some properties of this notion. The Alperin-Evens theorem (2.24.4(xiii)) is one of the main goals.

2.24.1 Definitions

Suppose X is a k-vector space graded by the non-negative integers. We say X has <u>growth</u> $\gamma(X) = \alpha$ provided α is the smallest non-negative integer such that there exists a constant μ with $\dim_k(X_n) \le \mu n^{\alpha-1}$ for all $n \ge 1$. If there is no such α we write $\gamma(X) = \infty$.

If V is a kG-module, let

$$..\ \to P_2 \to P_1 \to P_o \to V \to 0$$

be the minimal projective resolution of V. Namely P_o is the projective cover of V, P_1 is the projective cover of the kernel of $P_o \to V$, and so on. We define the <u>complexity</u> of V to be

$$cx(V) = cx_G(V) = \gamma(P_*).$$

The fact that $cx(V)$ is always finite will emerge in the course of the proof of 2.24.4, but of course easier arguments could be given if that was all we wanted to prove (cf. 2.22, property (vii) of cohomology and 2.31).

The p-rank of a group G is the rank of the largest elementary abelian subgroup of G.

A module V is periodic if for some $n \neq 0$, $V \cong \Omega^n V$. The least such n is called the period of V.

2.24.2 Lemma

Let G be a p-group and V a kG-module. Let $x\beta$ be the Bockstein of an element x of $H^1(G,\mathbb{Z}/p\mathbb{Z}) \subseteq H^1(G,k)$. Regarding x as an element of $[G/\Phi(G)]^*$, let H be the corresponding maximal subgroup of G. Then

$$\gamma(H^*(G,V)/H^*(G,V).(x\beta)) \leq \gamma(H^*(H,V{\downarrow_H})).$$

Proof

By 2.23.4, $H^{n-2}(G,V).(x\beta) = F^2H^n(G,V)$, and so by 2.22.5, $H^n(G,V)/(H^{n-2}(G,V).(x\beta))$ has a subspace isomorphic to $E_\infty^{1,n-1}(V)$ with quotient isomorphic to $E_\infty^{0,n}(V)$. Since G/H is cyclic, we have

$$\begin{aligned}\gamma(H^*(G,V)/H^*(G,V).(x\beta)) &= \gamma(E_\infty^{0,n}(V) \oplus E_\infty^{1,n-1}(V))\\ &\leq \gamma(H^0(G/H,H^n(H,V)) \oplus H^1(G/H,H^{n-1}(H,V)))\\ &\leq \gamma(H^n(H,V) \oplus H^{n-1}(H,V))\end{aligned}$$

(since $H^i(G/H,W)$ is a subquotient of W for any module W; see 2.22 exercise 2)

$$= \gamma(H^*(H,V)). \qquad \square$$

2.24.3 Lemma

Let G be a p-group, and let V be a kG-module. Then

(i) If H is a subgroup of index p in G then

$$\gamma(H^*(G,V)) \leq \gamma(H^*(H,V{\downarrow_H})) + 1$$

(ii) If G is not elementary abelian then

$$\gamma(H^*(G,V)) \leq \max_{H<G} \gamma(H^*(H,V{\downarrow_H})) .$$

Proof

(i) Let $\gamma(H^*(H,V{\downarrow_H})) = c$. Let $x\beta$ be the Bockstein of an element x of $H^1(G,k)$ corresponding to H. Then by 2.24.2,

$$\gamma(H^*(G,V)/H^*(G,V).(x\beta)) \leq \gamma(H^*(H,V{\downarrow_H})).$$

Hence for some constant λ,

$$\dim(H^n(G,V)/H^{n-2}(G,V).(x\beta)) \leq \lambda n^{c-1}$$

$$\dim(H^{n-2}(G,V).(x\beta)/H^{n-4}(G,V).(x\beta)^2) \leq \lambda(n-2)^{c-1}$$

and so on. Thus

$$\dim(H^n(G,V)) \le \lambda n^{c-1} + \lambda(n-2)^{c-1} + \lambda(n-4)^{c-1} + \dots$$
$$\le \lambda' n^c \quad \text{for some constant} \quad \lambda'.$$

(ii) Choose elements $x_1, \dots, x_r$ of $H^1(G,\mathbb{Z}/p\mathbb{Z}) \subseteq H^1(G,k)$ in accordance with Serre's theorem (2.23.3). Let H_i be the corresponding subgroups of G, and $c_i = \gamma(H^*(H_i, V{\downarrow}_{H_i}))$. Then since the $x_i\beta$ commute (see 2.22.1(iii)), 2.24.2 implies that for some constants $\lambda_1, \dots, \lambda_r$,

$$\dim(H^n(G,V)/H^{n-2}(G,V).(x_1\beta)) \le \lambda_1 n^{c_1-1}$$
$$\dim(H^{n-2}(G,V).(x_1\beta)/H^{n-4}(G,V).(x_1\beta)(x_2\beta))$$
$$\le \dim(H^{n-2}(G,V)/H^{n-4}(G,V).(x_2\beta)) \le \lambda_2(n-2)^{c_2-1}$$

and so on. Thus

$$\dim(H^n(G,V)) \le \lambda_1 n^{c_1-1} + \lambda_2(n-2)^{c_2-1} + \dots + \lambda_r(n-2r)^{c_r-1}$$
$$\le (\Sigma\lambda_i)\, n^{\max(c_i)-1}. \qquad \square$$

2.24.4 <u>Proposition</u>

(i) If $H \le G$ then $cx_H(V{\downarrow}_H) \le cx_G(V)$.

(ii) If $H \le G$ and W is a kH-module then $cx_G(W{\uparrow}^G) = cx_H(W)$.

(iii) $cx_G(V) = \max\limits_S \gamma(\mathrm{Ext}^*_G(V,S))$, where S runs over the simple kG-modules.

(iv) If G is a p-group, $cx_G(V) = \gamma(H^*(G,V^*))$, where V^* is the dual of V (in fact we shall see in 2.25.9 that $cx_G(V) = cx_G(V^*)$).

(v) If $0 \to V_1 \to V_2 \to V_3 \to 0$ is a short exact sequence of kG-modules then $cx_G(V_i) \le \max(cx_G(V_j), cx_G(V_k))$, $\{i,j,k\} = \{1,2,3\}$. In particular, the two largest complexities are equal.

(vi) $cx_G(V \oplus W) = \max(cx_G(V), cx_G(W))$.

(vii) If D is a vertex of V, then $cx_G(V) = cx_D(V{\downarrow}_D)$.

(viii) $cx_G(V \otimes W) \le \min(cx_G(V), cx_G(W))$

(ix) $cx_G(V) \le cx_G(k)$

(x) $cx_G(V) = 0$ if and only if V is projective.

(xi) $cx_G(V) = 1$ if and only if V is periodic $\oplus$ projective.

(xii) If G is a p-group and H is a subgroup of index p^n then $cx_H(V{\downarrow}_H) \le cx_G(V) \le cx_H(V{\downarrow}_H) + n$.

(xiii) $cx_G(V) = \max\limits_E cx_E(V{\downarrow}_E)$ as E ranges over the elementary abelian p-subgroups of G (Alperin, Evens).

(xiv) $cx_G(k)$ equals the p-rank of G.

(xv) $cx_G(V)$ is bounded by the p-rank of G for all V.

Proof

(i) A projective resolution of V is also a projective resolution of $V{\downarrow}_H$.

(ii) Inducing up a projective resolution of W to G gives a projective resolution of $W{\uparrow}^G$, with the property that the original resolution is a summand of the restriction (Mackey decomposition).

(iii) Let $S_1, \ldots, S_m$ be the simple kG-modules with projective covers $P_1, \ldots, P_m$. Let $d_i = \dim_k \mathrm{End}_{kG}(S_i)$. If $\ldots \to R_2 \to R_1 \to R_o \to V \to 0$ is a minimal projective resolution of V, then

$$\begin{aligned}\dim_k \mathrm{Ext}^n_G(V,S_i) &= \dim_k \mathrm{Hom}_{kG}(R_n,S_i)\\ &= d_i.(\text{multiplicity of } P_i \text{ as a summand of } R_n)\end{aligned}$$

Hence $\dim_k(R_n) = \Sigma(\dim(P_i)/d_i).\dim_k \mathrm{Ext}^n_G(V,S_i)$.

(iv) If G is a p-group, then k is the only simple kG-module, and so (iii) gives

$$\begin{aligned}cx_G(V) &= \gamma(\mathrm{Ext}^*_G(V,k))\\ &= \gamma(\mathrm{Ext}^*_G(k,V^*))\\ &= \gamma(H^*(G,V^*)) .\end{aligned}$$

(v) This follows from (iii) and the long exact Ext sequence.

(vi) This is clear by forming the direct sum of the resolutions.

(vii) $cx_G(V) \geq cx_D(V{\downarrow}_D) = cx_G(V{\downarrow}_D{\uparrow}^G) \geq cx_G(V)$, by (i), (ii) and (vi).

(viii) Tensoring a projective resolution of V with W gives a projective resolution of $V \otimes W$.

(ix) This follows from (viii) since $V \cong V \otimes k$.

(x) This is clear.

(xi) Assume without loss of generality that V has no projective direct summands. Let $\ldots \to P_1 \to P_o \to V \to 0$ be the minimal projective resolution of V. By 1.8.6 there is a homogeneous element x of positive degree j in $H^*(G,k)$ such that cup product with x induces an injection from $H^n(G,M^* \otimes S) \cong \mathrm{Ext}^n_G(M,S)$ to $\mathrm{Ext}^{n+j}_G(M,S)$, for n sufficiently large, and for each simple module S. Since $\gamma(\mathrm{Ext}^n_G(M,S)) \leq 1$, this injection is an isomorphism for n sufficiently large. On the chain level, cup product with x is represented by a map from P_n to P_{n+j}, which since the resolution is minimal (cf. the argument for (iii)), is an isomorphism commuting with the boundary homomorphism, for all n sufficiently large. Thus $\Omega^n M \cong \Omega^{n+j} M$ for some n, and so since M has no projective summands, this implies that $M \cong \Omega^j M$ (Eisenbud, [47]).

Remark

There is a much easier argument over a field which is algebraic over the ground field: pass down to a finite field, and remark that there are only finitely many modules of a given dimension.

(xii) By induction we need only prove this for $n = 1$. The first inequality follows from (i), while the second follows from (iv) and 2.24.3 (i).

(xiii) This follows from (i), (iv) and 2.24.3(ii).

(xiv) $cx_G(k) = \max_E cx_E(k)$ by (xiii).

$\leq$ p-rank of G by (xii).

Equality follows from the explicit structure of $H^*(E,k)$ given in 2.22 exercise 2.

(xv) This follows from (ix) and (xiv). □

2.24.5 Corollary (Chouinard)

A kG-module V is projective if and only if $V\downarrow_E$ is projective for every elementary abelian p-subgroup E of G.

Proof

This follows from 2.24.4(x) and (xiii). □

2.24.6 Corollary

A kG-module V is periodic if and only if $V\downarrow_E$ is periodic for every elementary abelian p-subgroup E of G.

Proof

This follows from 2.24.4(xi) and (xiii). □

2.24.7 Corollary

Let G be a p-group and V a kG-module whose restriction to some maximal subgroup is projective. Then V is periodic.

Proof

This follows from 2.24.4(x), (xi) and (xii). □

Example

In [64], Landrock and Michler examine the structure of the projective indecomposable modules for Janko's simple group J_1 over a splitting field of characteristic 2. It turns out that there is a simple module V of dimension 20 and a subgroup H of J_1 isomorphic to $L_2(11)$ such that $V{\downarrow}_H$ is projective. A Sylow 2-subgroup of H is contained to index two in a Sylow 2-subgroup of J_1 (which is in fact elementary abelian of order eight), and so by 2.24.7, V is periodic. In fact, there is a short exact sequence

$$0 \to \mho^3(V) \to P_1 \to \Omega^3(V) \to 0$$

and so V has period 7.

Exercises

1. Let G be a p-group and V a kG-module. Let $x\beta$ be the Bockstein of an element x of $H^1(G,\mathbb{Z}/p\mathbb{Z}) \subseteq H^1(G,k)$. Let H be the maximal subgroup of G corresponding to x, and let $F^pH^{p+q}(G,V)$ be the filtration associated with the spectral sequence

$$H^p(G/H,H^q(H,V)) \Rightarrow H^{p+q}(G,V).$$

(i) Using exercise 7 of 2.22, show that for $q \geq 1$,

$$\dim_k E_\infty^{p,q} \leq \dim_k E_2^{p,q} \leq |H|^{q+2} \dim_k H^1(H,V{\downarrow}_H).$$

(ii) Show that

$$\dim_k B_\infty^{n,o} \leq \sum_{r=2}^{n} \dim_k E_r^{n-r,r-1} \leq |H|^{n+2} \dim_k H^1(H,V{\downarrow}_H).$$

(iii) Let $U(n)$ denote the kernel of multiplication by $x\beta$ on $H^n(G,V)$. Using the fact that the map $b_2^{n,o}: E_2^{n,o} \to E_2^{n+2,o}$ given by cup product with $x\beta$ is an isomorphism for $n \geq 1$ (cohomology of cyclic groups), show that

$$\begin{aligned}\dim_k U(n) &\leq \dim_k(H^n(G,V)/F^nH^n(G,V)) + \dim_k(U(n) \cap F^nH^n(G,V)) \\ &\leq \sum_{r=0}^{n-1} \dim_k E_\infty^{r,n-r} + \dim_k B_\infty^{n+2,o} \\ &\leq 2|H|^{n+4}\dim_k H^1(H,V{\downarrow}_H).\end{aligned}$$

2. (Carlson [25])

Let G be a group and V a kG-module with vertex D. For a subgroup H of G, write $V{\downarrow}_H = W \oplus P$ where P is projective and W has no projective summands, and define $\mathrm{core}_H(V) = W$. Show that there is a constant B_G depending only on G such that if V has no projective summands then

$$\dim_k V \leq B_G \cdot \max_{\substack{E \leq D, \\ E \text{ elem.} \\ \text{abelian}}} \dim_k \mathrm{core}_E(V).$$

(Hint: first reduce to $G = D$ by the theory of vertices and sources, then use Serre's theorem 2.23.3 together with exercise 1(iii) and exercise 7(iii) of 2.22.)

Remark

This exercise may be used to give an alternative proof of the main results of this chapter, see Carlson [25].

2.25 Varieties associated to modules

In this section, we introduce an affine variety associated with a given kG-module V. This is a certain subvariety $X_G(V)$ of the spectrum $X_G = \mathrm{Max}(H^{ev}(G,k))$ of maximal ideals of the even cohomology ring. At this point it is appropriate to remark that if A is a commutative Noetherian graded ring, and A' is the subring generated by the A_i for i divisible by a given natural number n, then $\mathrm{Max}(A) \cong \mathrm{Max}(A')$, the isomorphism being given by $M \mapsto M \cap A'$. Thus it is fairly natural to pass down to the even cohomology, to obtain a commutative ring. It will turn out that the dimension of $X_G(V)$ is equal to the complexity of V.

In the case where G is elementary abelian, we show that $X_G(V)$ is naturally isomorphic to a variety $Y_G(V)$ defined in terms of the restriction of V to certain cyclic subgroups of kG.

In the next section, we shall see how the variety associated to a module for an arbitrary finite group is controlled by the restrictions to elementary abelian subgroups.

Throughout this and the next two sections, we assume that k is algebraically closed.

2.25.1 Definitions

Let $X_G = \mathrm{Max}(H^{ev}(G,k))$, the set of maximal ideals of $H^{ev}(G,k)$, as an affine variety with the Zariski topology. Since $H^{ev}(G,k)$ is a graded ring, we may also consider $\overline{X}_G = \mathrm{Proj}(H^{ev}(G,k))$, the set of homogeneous ideals, maximal in the ideal I of elements of positive degree. There is a natural morphism $X_G \setminus \{0\} \twoheadrightarrow \overline{X}_G$, where 0 is the point in X_G given by the ideal I. This homomorphism takes an ideal to the ideal generated by the homogeneous elements in it. We have $\dim(\overline{X}_G) = \dim(X_G) - 1$.

We denote by $\mathrm{Ann}_G(V)$ the ideal of $H^{ev}(G,k)$ consisting of those elements annihilating $H^*(G,V)$. The support of a module V, written $X_G(V)$, is the set of all maximal ideals $M \in X_G$ which contain $\mathrm{Ann}_G(V \otimes S)$ for some module S. We denote by $I_G(V)$ the ideal of $H^{ev}(G,k)$ consisting of those elements x such that for all modules S, there exists a positive integer j such that $H^*(G,V \otimes S).x^j = 0$. If $H \leq G$, then $H^*(H,V{\downarrow_H} \otimes S) \cong H^*(G,V \otimes S{\uparrow^G})$ by Shapiro's lemma, and so $res_{G,H}(I_G(V)) \subseteq I_H(V)$.

2.25.2 Lemma

Suppose $0 \to V' \to V \to V'' \to 0$ is a short exact sequence of kG-modules, and $x_1 \in Ann_G(V')$, $x_2 \in Ann_G(V'')$. Then $x_1x_2 \in Ann_G(V)$.

Proof

This follows from the long exact sequence of cohomology. □

Thus in the definition of $I_G(V)$, it is sufficient to check for S equal to the direct sum of the irreducible kG-modules.

2.25.3 Proposition

$X_G(V) = Max(H^{ev}(G,k)/I_G(V))$.

Proof

If $M \in X_G(V)$ then since M is prime, $I_G(V) \subseteq M$. Conversely, suppose $I_G(V) \subseteq M$. Let S be the direct sum of the irreducible kG-modules. Then $x \notin M$ implies $x^j \notin Ann_G(V \otimes S)$. Hence $M \in X_G(V)$. □

In particular, since $I_G(V)$ is a homogeneous ideal in $H^{ev}(G,k)$, $X_G(V)$ is a subvariety of X_G consisting of a union of lines through thr origin. Thus $\overline{X}_G(V) = Proj(H^{ev}(G,k)/I_G(V))$ is a projective subvariety of $\overline{X}_G$.

2.25.4 Proposition

$dim(X_G(V)) = cx_G(V^*)$.

Proof

Let S be the direct sum of the irreducible kG-modules. By 2.24.4(iii),

$$\begin{aligned} cx_G(V^*) &= \gamma(H^*(G,V \otimes S)) \\ &= \gamma(H^{ev}(G,V \otimes S)) \\ &= \gamma(H^{ev}(G,k)/I_G(V)) \quad \text{by 2.25.3} \\ &= dim(X_G(V)) \quad \text{by 1.8.7.} \end{aligned}$$ □

Remark

We shall see in 2.25.9 that $cx_G(V^*) = cx_G(V)$.

If $G = E$ is an elementary abelian p-group, there is another variety $Y_E(V)$ which we may associate to a module V. $J = J(kE)$ is a subspace of kE of codimension 1, and if $x \in J$, then $1 + x$ is an invertible element of kE of order p. Before defining $Y_E(V)$, we need proposition 2.25.6.

2.25.5 Lemma

Suppose V is a kE-module. If $x \in J$ has the property that $V{\downarrow}_{k\langle 1+x\rangle}$ is free, then a basis $e_1, \ldots, e_n$ of V may be found with the following properties.

(i) E acts upper triangularly (i.e. for $g \in E$, $e_i g \equiv e_i \bmod \langle e_{i+1}, \ldots, e_n\rangle$).

(ii) The element $1 + x \in kE$ is in Jordan canonical form.

Proof

Find a basis $f_1, \ldots, f_n$ of V such that $1 + x$ is in Jordan canonical form. Then the centralizer of $1 + x$ in $GL(V)$ consists of all matrices whose $p \times p$ blocks are of the form

$$\begin{pmatrix} \lambda_1 & \lambda_2 & & \cdots & \lambda_p \\ & \lambda_1 & \lambda_2 & \cdots & \\ & & \ddots & & \\ & 0 & & & \lambda_1 \end{pmatrix}$$

. Thus the collection of upper triangular matrices in this group form a Sylow p-subgroup, and so E may be conjugated into upper triangular form without disturbing $1 + x$. □

2.25.6 Proposition

Suppose V is a kG-module. If $x, y \in J$, and $x \equiv y \bmod J^2$, then $V{\downarrow}_{k\langle 1+x\rangle}$ is free if and only if $V{\downarrow}_{k\langle 1+y\rangle}$ is free.

Proof

Suppose $V{\downarrow}_{k\langle 1+x\rangle}$ is free. Choose a basis $e_1, \ldots, e_n$ for V as in 2.25.5. Then the entries on and immediately above the diagonal for y are the same as for x. Thus y^{p-1} has rank at least n/p (since the 1st, $(p+1)^{th}$, .., $(n-p+1)^{th}$ rows are linearly independent), so $V{\downarrow}_{k\langle 1+y\rangle}$ is free. □

For $V \neq 0$, we now define $Y_E(V)$ to be the subset of $Y_E = J/J^2$ consisting of zero together with the image in Y_E of the set of $x \in J$ such that $V{\downarrow}_{k\langle 1+x\rangle}$ is not free. For $V = 0$, we define $Y_E(V) = \emptyset$. Since $x \in Y_E(V)$ if and only if the rank of the matrix representing x^{p-1} is less than $\dim(V)/p$, $Y_E(V)$ is defined by polynomial equations (namely the vanishing of certain minors), and is hence a subvariety of Y_E.

The following theorem was conjectured by Carlson, and proved by Avrunin and Scott [9].

2.25.7 Theorem

There is a natural isomorphism $Y_E \cong X_E$, which has the property that for every module V, the image of $Y_E(V)$ is $X_E(V)$.

Remark

Before reading the proof, if the reader is not familiar with the cohomology of cyclic groups, he should turn back to 2.22, ex. 2.

Proof of 2.25.7

It is clear from 2.22 exercise 1 that $H^1(E,k)$ is naturally isomorphic to $(J/J^2)^*$. Now, the Bockstein homomorphism $\beta: H^1(E,k) \to H^2(E,k)$ is injective (see 2.23 and 2.22 exercise 2), and if $y_1, \dots, y_r$ form a basis for $H^1(E,k)$ then $\{y_i\beta\} \cup \{y_iy_j : i < j\}$ form a basis for $H^2(E,k)$. For p odd, the subalgebra of $H^{ev}(E,k)$ generated by $H^1(E,k)\beta$ forms a complement to $J(H^{ev}(E,k))$. For $p = 2$, $y\beta = y^2$ for each $y \in H^1(E,k)$. Thus in either case, $\mathrm{Max}(H^{ev}(E,k))$ is a vector space dual to $H^1(E,k)\beta \cong H^1(E,k) \cong (J/J^2)^*$, and hence X_E is naturally isomorphic to Y_E.

Now suppose $x \in J$. As remarked after 2.22.1, the map $k\langle 1+x\rangle \hookrightarrow kE$ gives rise to maps $H^*(E,k) \to H^*(\langle 1+x\rangle,k)$ and $H^*(E,V) \to H^*(\langle 1+x\rangle, V)$ commuting with cup products and the Bockstein map. Thus x determines a line through the origin in each of X_E and Y_E, and so we must check that one is in $X_E(V)$ if and only if the other is in $Y_E(V)$ (note that each of $X_E(V)$ and $Y_E(V)$ is a union of lines through the origin).

First, if the line determined by x in Y_E is in $Y_E(V)$, then $V{\downarrow}_{k\langle 1+x\rangle}$ is not free. Thus the support of $H^*(\langle 1+x\rangle, V)$ in $X_{\langle 1+x\rangle}$ is the whole of $X_{\langle 1+x\rangle}$, and so by the commutativity of the diagram

$$\begin{array}{ccc} H^{ev}(E,k) \times H^*(E,V) & \xrightarrow{\ \cup\ } & H^*(E,V) \\ \downarrow \text{res} & & \downarrow \text{res} \\ H^{ev}(\langle 1+x\rangle,k) \times H^*(\langle 1+x\rangle,V) & \xrightarrow{\ \cup\ } & H^*(\langle 1+x\rangle,V) \end{array}$$

it follows that the image of $X_{\langle 1+x\rangle}$ in X_G is in $X_G(V)$.

Conversely, suppose $V{\downarrow}_{k\langle 1+x\rangle}$ is free. Let F be a subgroup of kE containing $1 + x$, and isomorphic to E. The inclusion $F \hookrightarrow kE$ induces an isomorphism $kF \cong kE$. Thus we have a spectral sequence

$$H^p(F/\langle 1+x\rangle, H^q(\langle 1+x\rangle,V)) \Rightarrow H^{p+q}(F,V) = H^{p+q}(E,V),$$

and since $H^q(\langle 1+x\rangle,V) = 0$ for $q \neq 0$, this spectral sequence converges at the E_2 level. This implies that $H^*(F/\langle 1+x\rangle, V^{\langle 1+x\rangle}) \cong H^*(F,V)$, the isomorphism being given by the composite of the natural maps $H^*(F/\langle 1+x\rangle, V^{\langle 1+x\rangle}) \to H^*(F,V^{\langle 1+x\rangle}) \to H^*(F,V)$. Thus regarding $H^*(F,V)$ as an $H^*(F/\langle 1+x\rangle,k)$-module via the natural map $H^*(F/\langle 1+x\rangle,k) \to H^*(F,k)$, the above isomorphism is an $H^*(F/\langle 1+x\rangle,k)$-module isomorphism. In particular, Evens' theorem (see 2.22, property (vii) of cohomology) says that $H^{ev}(F/\langle 1+x\rangle, V^{\langle 1+x\rangle})$, and hence $H^{ev}(F,V)$, is finitely generated as an $H^*(F/\langle 1+x\rangle, k)$-module.

Now the composite map $H^*(F/\langle 1+x\rangle,k) \to H^*(F,k) \to H^*(\langle 1+x\rangle,k)$ sends all positive degree elements to zero. Letting P be the kernel of $H^*(F,k) \to H^*(\langle 1+x\rangle,k)$, we know that the points on the line $X_E = X_F$ corresponding to $\langle 1+x\rangle$ are the maximal ideals containing P. What we are asking is how many of these maximal ideals contain the annihilator of $H^{ev}(F,V)$. However, we know that $H^{ev}(F,V)/H^{ev}(F,V)P$ is a finitely generated module over $H^*(F/\langle 1+x\rangle,k)/\{\text{elts of positive degree}\} \cong k$, and is hence a finite dimensional vector space. In particular, this implies that only finitely many maximal ideals contain its annihilator, and so only finitely many points of $X_{\langle 1+x\rangle}$ belong to $X_E(V)$. Since k is an infinite field, this implies that $X_{\langle 1+x\rangle} \cap X_E(V) = \{0\}$. □

2.25.8 Corollary (Dade)

Let V be a module for an elementary abelian group E. Then V is free if and only if for every $x \in J\backslash J^2$, $V{\downarrow}_{k\langle 1+x\rangle}$ is free.

Proof

If V is free, it is clear that $V{\downarrow}_{k\langle 1+x\rangle}$ is free.

Conversely, suppose $V{\downarrow}_{k\langle 1+x\rangle}$ is free for all $x \in J\backslash J^2$. Then $Y_E(V) = 0$, and hence $X_E(V) = 0$ by 2.25.7. Thus by 2.25.4, $cx_E(V) = 0$, and so by 2.24.4(x), V is projective. □

2.25.9 Corollary (Carlson)

For any group G and any module V, $cx_G(V) = cx_G(V^*) = cx_G(V \otimes V^*) = \gamma(\mathrm{Ext}^n_G(V,V))$.

Proof

By 2.24.4 (iv) and (xiii), it is sufficient to prove this for G elementary abelian. In this case, $cx_G(V) = \dim(Y_G(V))$. But $V{\downarrow}_{k\langle 1+x\rangle}$ is free if and only if $V^*{\downarrow}_{k\langle 1+x\rangle}$ is free, which in turn happens if and only if $(V \otimes V^*){\downarrow}_{k\langle 1+x\rangle}$ is free, and so $Y_G(V) = Y_G(V^*) = Y_G(V \otimes V^*)$. Now use 2.25.4 and 2.25.7. □

2.25.10 Corollary

Let $E_1 \leq E_2$ be elementary abelian p-groups and V a kE_2-module. Identifying X_{E_1} with a subspace of X_{E_2} via $\mathrm{res}^*_{E_2,E_1}$, we have

$$X_{E_2}(V) \cap X_{E_1} = X_{E_1}(V). \quad \text{(see also 2.26.8)}$$

Proof

This follows immediately from 2.25.7, since it is clear from the definition that $Y_{E_2}(V) \cap Y_{E_1} = Y_{E_1}(V)$. □

Exercise

1. A kG-module V is said to be endotrivial if $V \otimes V^* \cong k \oplus P$, where k is the trivial module and P is projective. Using 2.24.5 and 2.18 exercise 3(ii), show that V is endotrivial if and only if $V{\downarrow_E}$ is endotrivial for every elementary abelian p-subgroup E of G.

Show that $V{\downarrow_E}$ is endotrivial if and only if

$$\dim_k \mathrm{End}_{kE}(V{\downarrow_E}) = 1 + (\dim_k V - 1)/|E|.$$

Deduce that $V{\downarrow_E}$ is endotrivial if and only if $V{\downarrow_{\langle 1+x\rangle}}$ is endotrivial for each $x \in J(kE)$. Show that if $x - y \in J^2(kE)$ then $V{\downarrow_{\langle 1+x\rangle}}$ is endotrivial if and only if $V{\downarrow_{\langle 1+y\rangle}}$ is endotrivial.

Remark

Dade [40] has shown that the only endotrivial modules for an abelian p-group are the modules $\Omega^n(k) \oplus$ projective and $\mho^n(k) \oplus$ projective $(n \geq 0)$.

Using this result, Puig has shown that for an arbitrary finite group, the multiplicative group of endotrivial modules (modulo projectives) is finitely generated.

2.26 The Quillen Stratification

We now investigate the variety $X_G(V)$ in relation to the elementary abelian subgroups of G. It turns out that it is a disjoint union of strata, one for each conjugacy class of elementary abelian subgroup E, and that each stratum is "homeomorphic to" a quotient of an affine variety determined by $V{\downarrow_E}$ by a regular group of automorphisms (see theorem 2.26.10)

2.26.1 Proposition

$$I_G(V) = \bigcap_{\substack{E \\ \text{elementary} \\ \text{abelian}}} \mathrm{res}^{-1}_{G,E}(I_E(V)) .$$

Proof

It is clear from the definitions that $I_G(V) \subseteq \bigcap_E \mathrm{res}^{-1}_{G,E}(I_E(V))$.

Let $P \in \mathrm{Syl}_p(G)$. Then $\mathrm{res}_{G,P}\colon H^*(G,k) \to H^*(P,k)$ is injective, and so $I_G(V) = \mathrm{res}^{-1}_{G,P}(I_P(V))$. Thus it is sufficient to prove the proposition for $G = P$ a p-group. We work by induction on $|P|$. If P is elementary abelian, the proposition is clear, so assume P is not elementary abelian. Choose elements $x_1, \dots, x_r \in H^1(P,\mathbb{Z}/p\mathbb{Z})$ in accordance with Serre's theorem (2.23.3), and let $P_1, \dots, P_r$ be the corresponding maximal subgroups of P.

For each x_i, we have a spectral sequence

$$H^p(P/P_i, H^q(P_i,V)) \Rightarrow H^{p+q}(P,V)$$

and by the Quillen-Venkov lemma (2.23.4), $F^2H^{n+2}(P,V) = H^n(P,V)(x_i\beta)$. Suppose $x \in H^*(P,k)$, such that for all i, $\mathrm{res}_{P,P_i}(x) \in I_{P_i}(V)$. We must show that $x \in I_P(V)$. We have $\mathrm{res}_{P,P_i}(x) \in H^*(P_i,k)^P = H^o(P/P_i, H^*(P_i,k)) = E_2^{o,*}$. Now $E_2^{o,*}$ acts on $E_2^{p,*}(V)$, and for some j independent of p and q, x annihilates each $E_2^{p,q}(V)$ (since $H^p(P/P_i, H^q(P_i,V))$ is a subquotient of $H^q(P_i,V)$, see 2.22 exercise 2). Since $x \in H^*(P,k)$, $\mathrm{res}_{P,P_i}(x)$ is an element of $E_2^{o,*}$ which survives to the E_∞ level, and at each stage the jth power annihilates each $E_r^{p,q}(V)$. Thus at the E_∞ level, x^j annihilates each $E_\infty^{p,q}(V)$, and so $H^*(P,V)x^{2j} \subseteq F^2H^*(P, V) = H^*(P,V)(x_i\beta)$. Thus $H^*(P,V)x^{2jr} \subseteq H^*(P,V)(x_1\beta)\dots(x_r\beta) = 0$. □

If H is a subgroup of G, we have a map $t_{H,G} = \mathrm{res}^*_{G,H}: X_H \to X_G$. It is clear that $t_{H,G}(X_H(V)) \subseteq X_G(V)$ since $\mathrm{res}_{G,H}(I_G(V)) \subseteq I_H(V)$. One of our goals in this section will be theorem 2.26.8, which is a sort of converse to this.

2.26.2 <u>Corollary</u>

$$X_G(V) = \bigcup_{\substack{E \\ \text{elementary} \\ \text{abelian}}} t_{E,G}(X_E(V)) .$$

<u>Proof</u>

Clear. □

<u>Remark</u>

2.26.2 and 2.25.7 enable us to reduce questions about $X_G(V)$ to questions about cyclic subgroups of the group algebra of order p.

2.26.3 <u>Corollary</u>

$$X_G = \bigcup_{\substack{E \\ \text{elementary} \\ \text{abelian}}} t_{E,G}(X_E) . \qquad \square$$

2.26.4 Corollary

An element $x \in H^i(G,k)$ is nilpotent if and only if $\mathrm{res}_{G,E}(x)$ is nilpotent for all elementary abelian subgroups E of G.

Proof

This is the case $V = k$ of 2.26.1. □

For E an elementary abelian p-group of rank r, we know that X_E is a vector space of dimension r. We define

$$X_E^+ = X_E \setminus \bigcup_{E'< E} t_{E',E}(X_{E'})$$

$$X_{G,E} = t_{E,G}(X_E), \qquad X_{G,E}^+ = t_{E,G}(X_E^+)$$

$$X_E^+(V) = X_E(V) \setminus \bigcup_{E'< E} t_{E',E}(X_{E'}(V))$$

$$X_{G,E}(V) = t_{E,G}(X_E(V)) \ , \qquad X_{G,E}^+(V) = t_{E,G}(X_E^+(V)) .$$

Thus X_E^+ is the space X_E with all the hyperplanes defined over $\mathbb{Z}/p\mathbb{Z}$ removed.

Let $\sigma_E = \prod_{x \in H^1(E,\mathbb{Z}/p\mathbb{Z})} (x\beta)$. Then σ_E may be regarded as an element of the coordinate ring of X_E, and the open set defined by σ_E is X_E^+. Thus the coordinate ring of the variety X_E^+ is $H^{ev}(E,k)[\sigma_E^{-1}]$.

We now use the norm map (see 2.22, properties of cohomology (viii)) to ensure that $H^{ev}(G,k)$ has enough elements.

2.26.5 Lemma

Let E be an elementary abelian p-subgroup of G, and let $|N_G(E):E| = p^a.h$ with $(p,h) = 1$. Then

(i) There exists an element ρ_E of $H^{ev}(G,k)$ such that for elementary abelian p-subgroups E' of G,

$$\mathrm{res}_{G,E'}(\rho_E) = \begin{cases} 0 & \text{if } E \text{ is not conjugate to a subgroup of } E' \\ (\sigma_E)^{p^a} & \text{if } E = E' . \end{cases}$$

(ii) If $y \in H^{ev}(E,k)$ is invariant under $N_G(E)$ then there is an element $y' \in H^{ev}(G,k)$ with

$$\mathrm{res}_{G,E}(y') = (y\,\sigma_E)^{p^a} .$$

Proof

(i) Let $z = \mathrm{norm}_{E,G}(1+\sigma_E)$. The Mackey formula gives

$$\begin{aligned}\mathrm{res}_{G,E}(z) &= (1+\sigma_E)^{p^a.h} \\ &= 1 + h(\sigma_E)^{p^a} + \text{terms of higher degree.}\end{aligned}$$

Also, if E is not conjugate to a subgroup of E', then $\mathrm{res}_{G,E}(z) = 1$, since for any proper subgroup E'' of E, $\mathrm{res}_{E,E''}(\sigma_E) = 0$. Thus we may take $(1/h)$ times the homogeneous part of z of degree $p^a.\deg(\sigma_E)$ as our ρ_E.

(ii) Suppose without loss of generality that y is homogeneous. Let $z' = \mathrm{norm}_{E,G}(1 + y\,\sigma_E)$. Then

$$\mathrm{res}_{G,E}(z') = 1 + h(y\,\sigma_E)^{p^a} + \text{terms of higher degree.}$$

Thus we may take $(1/h)$ times the homogeneous part of z' of degree $p^a.\deg(y\,\sigma_E)$ as our y'. □

2.26.6 Definitions

Given $x \in X_G$, 2.26.3 shows that there exists an elementary abelian p-subgroup $E \le G$ and $y \in X_E$ such that $x = t_{E,G}(y)$. If such a pair (E,y) satisfies the minimality condition that there does not exist a subgroup $E' < E$ and $z \in X_{E'}$ with $y = t_{E',E}(z)$ we say that E is a vertex of x and y is a source (by analogy with section 2.5).

2.26.7 Theorem

(i) Given $x \in X_G$, suppose (E_1,y_1) and (E_2,y_2) are both vertices and sources for x. Then there is an element $g \in G$ with $E_1{}^g = E_2$ and $y_1{}^g = y_2$.

(ii) (Quillen stratification of X_G)

X_G is a disjoint union of the locally closed subvarieties $X^+_{G,E}$, as E runs over a set of representatives of conjugacy classes of elementary abelian p-subgroups of G. The group $W_G(E) = N_G(E)/C_G(E)$ acts freely on X^+_E, and $t_{E,G}$ induces a finite homeomorphism

$$X^+_E/W_G(E) \to X^+_{G,E} \quad .$$

The topology on X_G is given as follows. The natural map $\varinjlim_E X_E \to X_G$ is a finite homeomorphism. (The morphisms in the limit symbol are compositions of inclusions and conjugations).

Note

The finite homeomorphism $X^+_E/W_G(E) \to X^+_{G,E}$ is called by Quillen an "inseparable isogeny", since it means that at the level of coordinate rings, there are inclusions

$$k[X_E^+/W_G(E)] \supseteq k[X_{G,E}^+] \supseteq k[X_E^+/W_G(E)]^{p^a}$$

where p^a is the power of p appearing in 2.26.5. In fact the argument can be strengthened to show that p^a may be taken to be the p-part of $|C_G(E):E|$, see [9].

<u>Proof of 2.26.7</u>

For each elementary abelian subgroup $E \leq G$, we have a map $X_E^+ \to X_{G,E}^+$. Lemma 2.26.5 shows that for the corresponding map of coordinate rings

$$\begin{array}{ccc} k[X_{G,E}^+] & \supseteq & k[X_{G,E}][\rho_E^{-1}] \\ \downarrow & & \\ k[X_E^+] & \cong & H^{ev}(E,k)[\sigma_E^{-1}] \end{array} ,$$

the p^a-th power of any element of $k(X_E^+)$ invariant under $W_G(E)$ is in the image. This means that if we look at the extensions of function fields

$$k(X_{G,E}^+) \hookrightarrow k(X_E^+)^{W_G(E)} \hookrightarrow k(X_E^+) ,$$

the first extension is purely inseparable, while the second is Galois, with Galois group $W_G(E)$. Thus the map $X_E^+/W_G(E) \to X_{G,E}^+$ is a finite homeomorphism (or inseparable isogeny).

Next, if E_1 is not conjugate to a subgroup of E_2, then by 2.26.5, ρ_{E_1} is invertible on X_{G,E_1}^+ and zero on X_{G,E_2}^+. Thus the different $X_{G,E}^+$ are disjoint for non-conjugate E's. Moreover, 2.26.3 shows that the union of the $X_{G,E}^+$ is X_G. This completes the proof of (i) and the first part of (ii).

It now remains to study the glueing together of the $X_{G,E}^+$ to form X_G. Since there are only finitely many elementary abelian p-subgroups of G, and for each one the map $X_E \to X_{G,E}$ is finite, it follows that the map $\varinjlim_E X_E \to X_G$ is finite. The bijectivity follows from the fact that the $X_{G,E}^+$ are disjoint. □

<u>Example</u>

R. Wilson [94] has shown that Lyons' simple group Ly has exactly two conjugacy classes of maximal elementary abelian subgroups, of order 2^3 and 2^4, with normalizers $2^3.L_3(2)$ and $2^4.3A_7$. These may be chosen to intersect in a subgroup of order 2^2. Thus X_G has two irreducible components, of dimension three and four, and their intersection has dimension two.

2.26.8 Theorem

Let H be a subgroup of G, and V a kG-module. Then

$$X_H(V) = t_{H,G}^{-1}(X_G(V)).$$

Proof

It is clear that $t_{H,G}(X_H(V)) \subseteq X_G(V)$, so it remains to show that $t_{H,G}^{-1}(X_G(V)) \subseteq X_H(V)$. Let $x \in t_{H,G}^{-1}(X_G(V))$. By 2.26.2 we may choose $E_1 \leq G$, and $y_1 \in X_{E_1}(V)$, with $t_{H,G}(x) = t_{E,G}(y_1)$. Let (E_2,y_2) be a vertex and source of x. Then (E_2,y_2) are also a vertex and source of $t_{H,G}(x)$, and so by 2.26.7(i), there exists $g \in G$ with $E_2{}^g \leq E_1$ and $t_{E_2{}^g,E_1}(y_2{}^g) = y_1$. By 2.25.10, it follows that $y_2{}^g \in X_{E_2{}^g}(V)$, and so $y_2 \in X_{E_2}(V)$, and $x = t_{E_2,H}(y_2) \in X_H(V)$. □

The following theorem summarizes some of the main properties of the cohomology varieties. See also 2.26.10, 2.27 and 2.28.7.

2.26.9 Theorem

Let $H \leq G$, let V be a kG-module and W a kH-module.

(i) $\dim(X_G(V)) = cx_G(V)$

(ii) $X_G(V) = X_G(V^*) = X_G(V \otimes V^*) = X_G(\Omega^n V)$

(iii) If $0 \to V_1 \to V_2 \to V_3 \to 0$ is a short exact sequence of kG-modules then $X_G(V_i) \subseteq X_G(V_j) \cup X_G(V_k)$ for $\{i,j,k\} = \{1,2,3\}$.

(iv) $X_G(V \oplus V') = X_G(V) \cup X_G(V')$

(v) $X_G(V \otimes V') = X_G(V) \cap X_G(V')$

(vi) $X_H(V{\downarrow_H}) = t_{H,G}^{-1}(X_G(V))$

(vii) $X_G(W{\uparrow^G}) = t_{H,G}(X_H(W))$

(viii) $X_G(V) = \{0\}$ if and only if V is projective

(ix) $X_G(V) = \bigcup_E t_{E,G}(X_E(V{\downarrow_E}))$ as E ranges over the set of elementary abelian p-subgroups of G.

(x) If (p) denotes the Frobenius map on both modules and varieties than $X_G(V^{(p)}) = X_G(V)^{(p)}$.

Proof

(i) See 2.25.4 and 2.25.9.

(ii) By 2.26.2 it suffices to prove these equalities for $G = E$ elementary abelian. Thus by 2.25.7 we must show that

$$Y_E(V) = Y_E(V^*) = Y_E(V \otimes V^*) = Y_E(\Omega^n V).$$

But this follows from 1.4.4 and the definition of $Y_E(V)$.

(iii) This follows in a similar way from 2.25.7 and 2.26.2.

(iv) This is clear.

(v) Consider $U \otimes V$ as a $k(G \times G)$-module. Then the Künneth formula shows that we have

$$\begin{array}{ccc} X_{G\times G}(V \otimes V') & \hookrightarrow & X_{G\times G} \\ \wr\!\| & & \| \\ X_G(V) \times X_G(V') & \hookrightarrow & X_G \times X_G . \end{array}$$

The diagonal map $G \hookrightarrow G \times G$ gives rise to the diagonal map $X_G \hookrightarrow X_G \times X_G$. Hence by 2.26.8,

$$\begin{aligned} X_G(V \otimes V') &= t^{-1}_{G,G\times G}(X_{G\times G}(V \otimes V')) \\ &= t^{-1}_{G,G\times G}(X_G(V) \times X_G(V')) \\ &= X_G(V) \cap X_G(V'). \end{aligned}$$

(vi) See 2.26.8.

(vii) Since $W{\uparrow}^G{\downarrow}_H$ has W as a direct summand,

$$t_{H,G}(X_H(W)) \subseteq t_{H,G}(X_H(W{\uparrow}^G{\downarrow}_H)) \subseteq X_G(W{\uparrow}^G).$$

Conversely if $x \in X_G(W{\uparrow}^G)$, let (E,y) be a vertex and source of x. Then by (vi), (iv) and Mackey decomposition,

$$y \in X_E(W{\uparrow}^G{\downarrow}_E) = X_E(\sum_{HgE} W^g{\downarrow}_{H^g\cap E}{\uparrow}^E) = \bigcup_{HgE} X_E(W^g{\downarrow}_{H^g\cap E}{\uparrow}^E).$$ Thus

replacing (E,y) by a conjugate if necessary, we have

$$y \in X_E(W{\downarrow}_{H\cap E}{\uparrow}^E) = Y_E(W{\downarrow}_{H\cap E}{\uparrow}^E)$$ by 2.25.7, and so $E \leq H$

by minimality of E. Now since $y \in X_E(W{\downarrow}_E)$, $t_{E,H}(y) \in X_H(W)$ and $x = t_{H,G}(t_{E,H}(y)) \in t_{H,G}(X_H(W))$.

(viii) This follows from (i) and 2.24.4(x).

(ix) See 2.26.2.

(x) The Frobenius map acts on the bar resolution $\tilde{X}_i(G)$ and induces an isomorphism $\mathrm{Hom}(\tilde{X}_i(G), V^{(p)}) \cong \mathrm{Hom}(\tilde{X}_i(G),V)^{(p)}$ commuting with the δ's. □

2.26.10 <u>Corollary</u>

Suppose V and V' are modules and $\overline{X}_G(V) \cap \overline{X}_G(V') = \emptyset$ (i.e. $X_G(V) \cap X_G(V') = \{0\}$). Then $\mathrm{Ext}^i_G(V,V') = 0$ for all $i > 0$.

Proof

$X_G(V^* \otimes V') = X_G(V) \cap X_G(V') = \{0\}$ by 2.26.9(ii) and (v), and so by (viii), $V^* \otimes V'$ is projective. Thus

$$\mathrm{Ext}^i_G(V,V') \cong H^i(G,V^* \otimes V') = 0. \quad \square$$

Remark

It follows from 2.26.9 that if X is a subset of $\overline{X}_G$ then the linear span $A(G,X) \subseteq A_k(G)$ of modules V for which $\overline{X}_G(V) \subseteq X$, is an ideal in $A_k(G)$. We shall obtain some information about these ideals in 2.27.9.

Finally, we have the module analogue of 2.26.6(ii).

2.26.11 Theorem (Quillen stratification of $X_G(V)$, Avrunin-Scott)

$X_G(V)$ is a disjoint union of the locally closed subvarieties $X^+_{G,E}(V)$, as E runs over a set of representatives of conjugacy classes of elementary abelian p-subgroups of G. The group $W_G(E) = N_G(E)/C_G(E)$ acts freely on $X^+_E(V)$, and $t_{E,G}$ induces a finite homeomorphism

$$X^+_E(V)/W_G(E) \to X^+_{G,E}(V).$$

The natural map

$$\lim_{\vec{E}} X_E(V) \to X_G(V)$$

is a bijective finite morphism.

Proof

This follows from 2.26.6(ii) and 2.26.8. $\square$

Note that $X^+_E(V)$ and $X^+_{G,E}(V)$ are empty unless E is contained in a vertex of some direct summand of V.

Exercise

If X is a projective variety of dimension d in $\mathbb{P}^n$ then there is a linear subspace of dimension $n-d-1$ not intersecting X. Use this fact to show that if V is a kG-module and

$$s = (\text{p-rank of } G) - cx_G(V)$$

then $p^s \mid \dim_k(V)$.

(Hint: first restrict to a suitable elementary abelian subgroup E of G, and then restrict to a subgroup of kE of order p^s to obtain a projective module).

2.27 What varieties can occur?

We know from section 2.25 that for every kG-module V we have a subvariety $X_G(V)$ of the variety $X_G = \mathrm{Max}(H^{ev}(G,k))$. In this section we investigate the following questions.

(1) Which subvarieties of X_G occur as $X_G(V)$ for some module V?

(2) Which subvarieties of X_G occur as $X_G(V)$ for some indecomposable module V?

We obtain a complete and simple answer to (1), namely that every homogeneous subvariety of X_G is of the form $X_G(V)$ for some module V (2.27.3).

A partial answer to (2) is given in Carlson's result 2.27.8, which states that if V is indecomposable then $\overline{X}_G(V)$ is topologically connected.

We shall see that every irreducible homogeneous subvariety of X_G occurs as $X_G(V)$ for some indecomposable module V, but the question of exactly which connected non-irreducible varieties can occur is still open.

The following construction for modules L_ζ is basic to the ensuing discussion, and lemma 2.27.2 is what allows us to prove results by induction on dimension.

For any kG-module V, there is a natural isomorphism $\mathrm{Ext}^n_G(V,V) \cong (\Omega^n V, V)^{1,G}$ by 2.19.1. Thus an element $\zeta \in \mathrm{Ext}^n_G(V,V)$ is represented by a homomorphism $\Omega^n V \to V$, and such a homomorphism represents the zero element if and only if it factors through a projective module. In particular, for $V = k$, the trivial kG-module, no homomorphisms $\Omega^n k \to k$ factor through projective modules, and so for each $\zeta \in H^n(G,k) \cong \mathrm{Ext}^n_G(k,k)$ we have a well defined homomorphism $\Omega^n k \to k$ whose kernel we denote by L_ζ. Thus we have a short exact sequence

$$0 \longrightarrow L_\zeta \longrightarrow \Omega^n k \xrightarrow{\zeta} k \longrightarrow 0.$$

2.27.1 Lemma

Let ϕ be the natural homomorphism from $\mathrm{Ext}^n_G(k,k)$ to $\mathrm{Ext}^n_G(V,V)$ given by tensoring with V. If $\zeta \in \mathrm{Ker}(\phi)$ then

$$L_\zeta \otimes V \cong \Omega^n V \oplus \Omega V \oplus P$$

where P is a projective module.

Proof

We tensor the short exact sequence defining L_ζ with V to give

$$0 \longrightarrow L_\zeta \otimes V \longrightarrow \Omega^n k \otimes V \xrightarrow{\zeta\otimes 1} V \longrightarrow 0.$$

Since $\zeta \in \mathrm{Ker}(\phi)$, $\zeta \otimes 1$ factors through a projective module. Thus by 1.4.2 (i),

$$L_\zeta \otimes V \oplus P_V \cong \Omega^n k \otimes V \oplus \Omega V$$

where P_V is the projective cover of V. Now $\Omega^n k \otimes V \cong \Omega^n V \oplus$ projective by 1.4.2 (ii), and so the result follows from the Krull-Schmidt theorem. □

2.27.2 Lemma

For ζ a homogeneous element of $H^{ev}(G,k)$, $X_G(L_\zeta)$ is the hypersurface $X_G(\zeta)$ defined by ζ considered as an element of the coordinate ring of X_G.

Proof

By 2.26.2, $X_G(L_\zeta) = \bigcup_{E \text{ elem. abelian}} t_{E,G}(X_E(L_\zeta\downarrow_E))$, and by 2.26.3 $X_G(\zeta) = \bigcup_{E \text{ elem. abelian}} t_{E,G}(X_E(\mathrm{res}_{G,E}(\zeta)))$. It thus suffices to prove the lemma in the case where $G = E$ is elementary abelian. Following through the identification of X_E with Y_E given in the proof of 2.25.7, we see that we are required to prove that

$$Y_E(L_\zeta) = \{0\} \cup \{\bar{x} \in J/J^2 : \mathrm{res}_{E,<1+x>}(\zeta) = 0\}.$$

But by Schanuel's lemma (1.4.2 (ii)), $\Omega^n(k)\downarrow_{<1+x>} \cong k \oplus P$ with P projective (for a cyclic group of order p, $\Omega^n(k) \cong k$ since n is even). Thus $\mathrm{res}_{E,<1+x>}(\zeta) = 0$ if and only if the map $k \oplus P \to k$ in the short exact sequence

$$0 \longrightarrow L_\zeta\downarrow_{<1+x>} \longrightarrow k \oplus P \longrightarrow k \longrightarrow 0$$

factors through a projective module. If it does factor then $L_\zeta\downarrow_{<1+x>} \cong k \oplus \Omega(k) \oplus$ projective, while if it does not factor then $L_\zeta\downarrow_{<1+x>} \cong P$, and so we are done. □

2.27.3 Theorem (Carlson)

Every homogeneous subvariety of X_G is of the form $X_G(V)$ for some module V.

Proof

Suppose X is a homogeneous subvariety of X_G, and the corresponding ideal in $H^{ev}(G,k)$ is I. Since $H^{ev}(G,k)$ is

Noetherian (see 2.22, properties of cohomology (vii)), I is finitely generated by homogeneous elements $I = \langle \zeta_1, \ldots, \zeta_r \rangle$. Then by 2.26.9(ii) and 2.27.2,

$$\begin{aligned} X_G(L_{\zeta_1} \otimes \ldots \otimes L_{\zeta_r}) &= X_G(L_{\zeta_1}) \cap \ldots \cap X_G(L_{\zeta_r}) \\ &= X_G(\zeta_1) \cap \ldots \cap X_G(\zeta_r) \\ &= X. \qquad \square \end{aligned}$$

2.27.4 Corollary

The map $X \mapsto A(G,X)$ (see remark after 2.26.9) is an inclusion preserving injection from the set of subsets of $\overline{X}_G$ to the set of ideals of $A_k(G)$. $\square$

2.27.5 Corollary

Every irreducible homogeneous subvariety of X_G is of the form $X_G(V)$ for some indecomposable module V.

Proof

This follows from 2.27.3 and 2.26.9 (iv). $\square$

2.27.6 Lemma

Suppose ζ_1 and ζ_2 are homogeneous elements of $H^{ev}(G,k)$ of degrees r and s respectively. Then there is a short exact sequence $0 \to \Omega^r(L_{\zeta_2}) \to L_{\zeta_1\zeta_2} \oplus P \to L_{\zeta_1} \to 0$ with P a projective module.

Proof

Tensor the short exact sequence $0 \to L_{\zeta_2} \to \Omega^s(k) \xrightarrow{\zeta_2} k \to 0$ with $\Omega^r(k)$ to obtain a short exact sequence

$$0 \to \Omega^r(L_{\zeta_2}) \oplus \text{projective} \to \Omega^{r+s}(k) \oplus \text{projective} \to \Omega^s(k) \to 0 .$$

Since projectives are injective (1.4.4), we may subtract out projective modules from the first two terms to obtain

$$0 \to \Omega^r(L_{\zeta_2}) \to \Omega^{r+s}(k) \oplus P \to \Omega^s(k) \to 0.$$

We now form the pullback diagram

$$\begin{array}{ccccccccc}
 & & & & 0 & & 0 & & \\
 & & & & \downarrow & & \downarrow & & \\
0 & \longrightarrow & \Omega^r(L_{\zeta_2}) & \longrightarrow & L_{\zeta_1\zeta_2} \oplus P & \longrightarrow & L_{\zeta_1} & \longrightarrow & 0 \\
 & & \| & & \downarrow & & \downarrow & & \\
0 & \longrightarrow & \Omega^r(L_{\zeta_2}) & \longrightarrow & \Omega^{r+s}(k) \oplus P & \longrightarrow & \Omega^s(k) & \longrightarrow & 0 \\
 & & & & \downarrow{\scriptstyle \zeta_1\zeta_2} & & \downarrow{\scriptstyle \zeta_1} & & \\
 & & & & k & = & k & & \\
 & & & & \downarrow & & \downarrow & & \\
 & & & & 0 & & 0 & . &
\end{array}$$ □

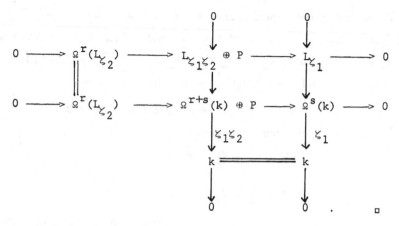

2.27.7 <u>Theorem</u> (Carlson).

If $X_G(V) \subseteq X_1 \cup X_2$, where X_1 and X_2 are homogeneous subvarieties of X_G with $X_1 \cap X_2 = \{0\}$, then we may write $V = V_1 \oplus V_2$ with $X_G(V_1) \subseteq X_1$ and $X_G(V_2) \subseteq X_2$.

<u>Proof</u>

We prove this by induction on $d = \dim(X_1) + \dim(X_2)$. The result is clear when $d = 0$ or 1, so suppose $d > 1$. Choose ζ_1 and ζ_2 homogeneous elements of $H^{ev}(G,k)$ of degrees r and s respectively, with

(i) $X_1 \subseteq X_G(\zeta_1)$ and $\dim(X_2 \cap X_G(\zeta_2)) = \dim(X_2) - 1$, and

(ii) $X_2 \subseteq X_G(\zeta_2)$ and $\dim(X_1 \cap X_G(\zeta_1)) = \dim(X_1) - 1$.

Then $X_G(V) \subseteq X_1 \cup X_2 \subseteq X_G(\zeta_1) \cup X_G(\zeta_2) = X_G(\zeta_1\zeta_2)$ and so $\zeta_1\zeta_2 \in I_G(V)$. Thus replacing ζ_1 and ζ_2 by suitable powers, we may assume that $\zeta_1\zeta_2 \in \operatorname{Ker}(\phi)$, where ϕ is the natural map from $\operatorname{Ext}^*_G(k,k)$ to $\operatorname{Ext}^*_G(V,V) = H^*(G, V \otimes V^*)$. Thus by 2.27.1

$$L_{\zeta_1\zeta_2} \otimes V \cong \Omega^{r+s}(V) \oplus \Omega(V) \oplus \text{projective}.$$

Now tensor V with the short exact sequence given in 2.27.6 to obtain

$$0 \to \Omega^r(L_{\zeta_2}) \otimes V \to \Omega^{r+s}(V) \oplus \Omega(V) \oplus \text{projective} \to L_{\zeta_1} \otimes V \to 0.$$

Now by 2.26.9 and 2.27.2,

$$X_G(L_{\zeta_1} \otimes V) = X_G(V) \cap X_G(\zeta_1)$$

and

$$X_G(\Omega^r(L_{\zeta_2}) \otimes V) = X_G(V) \cap X_G(\zeta_2).$$

Thus by the inductive hypothesis, $L_{\zeta_1} \otimes V = W_1 \oplus W_2$ with $X_G(W_1) \subseteq X_1 \cap X_G(\zeta_1)$ and $X_G(W_2) \subseteq X_2$, and $\Omega^r(L_{\zeta_2}) \otimes V = W_1' \oplus W_2'$ with $X_G(W_1') \subseteq X_1$ and $X_G(W_2') \subseteq X_2 \cap X_G(\zeta_2)$. Now by 2.26.10, $\mathrm{Ext}^1_G(W_1, W_2') = 0$ and $\mathrm{Ext}^1_G(W_2, W_1') = 0$. Thus we have

$$\Omega^n(V) \oplus \Omega(V) \oplus \text{projective} = V_1 \oplus V_2$$

where there are short exact sequences

$$0 \to W_1' \to V_1 \to W_1 \to 0$$

and

$$0 \to W_2' \to V_2 \to W_2 \to 0 .$$

The result now follows from 2.26.9(ii) and the Krull-Schmidt theorem. □

2.27.8 Corollary

If V is indecomposable then $\overline{X}_G(V)$ is topologically connected (in the Zariski topology). □

We now have enough information to state the main properties of the ideals $A(G,X)$ introduced after 2.26.10 (see also 2.28.8).

2.27.9 Theorem

Let $H \leq G$, let X be a subset of $\overline{X}_G$ and X' a subset of $\overline{X}_H$.

(i) $A(G,X)$ is an ideal in $A(G)$

(ii) $A(H, t^{-1}_{H,G}(X)) \supseteq r_{G,H}(A(G,X))$

(iii) $A(G, t_{H,G}(X')) \supseteq i_{H,G}(A(H,X'))$

(iv) $A(G,X)$ is closed under taking dual modules and under Ω.

(v) If $V \otimes V^* \in A(G,X)$ then so are V and V^*.

(vi) $A(G,X)$ is closed under taking extensions of modules.

(vii) $A(G,\emptyset) = A(G,1)$, the linear span of the projective modules.

(viii) $A(G, X_1 \cap X_2) = A(G,X_1) \cap A(G,X_2)$

(ix) $A(G, X_1 \cup X_2) \supseteq A(G,X_1) + A(G,X_2)$ with equality if $X_1 \cap X_2 = \emptyset$.

(x) If $X_1 \subsetneq X_2$ then $A(G,X_1) \subsetneq A(G,X_2)$

(xi) $\psi^n(A(G,X)) \subseteq A(G,X)$ for n coprime to p (see 2.16), while $\psi^p(A(G,X)) \subseteq A(G,X^{(p)})$, where ${}^{(p)}$ is the Frobenius map on varieties. Thus if $X = X^{(p)}$, $A(G,X)$ is stable under the operations λ^n.

Proof

(i) See 2.26.9(iv) and (v).

(ii) See 2.26.9(vi).

(iii) See 2.26.9(vii).

(iv) See 2.26.9(ii).

(v) See 2.26.9(ii)

(vi) See 2.26.9(iii), $i = 2$.

(vii) See 2.26.9(viii).

(viii) Clear.

(ix) This follows from 2.27.7.

(x) See 2.27.4.

(xi) Suppose $X_G(V) \subseteq X$. Then for n coprime to p, $\psi^n(V)$ is a linear combination of direct summands of $\otimes^n(V)$, and hence lies in $A(G,X)$. For $n = p$, 2.26.9(x) tells us that $\psi^p(A(G,X)) \subseteq A(G,X^{(p)})$. Finally, the λ^n are linear combinations of the ψ^n. □

Note that it is difficult to make a ring theoretic statement corresponding to 2.26.9(ix). It is tempting to write $A(G,X) = \bigcap_E r^{-1}_{G,E}(A(E,t^{-1}_{E,G}(X)))$, but this is false in general. For example, the right hand side contains $V_2 - V_1 - V_3$ whenever $0 \to V_1 \to V_2 \to V_3 \to 0$ is a short exact sequence which splits on restriction to every elementary abelian p-subgroup of G, and we know that there are plenty of these by 2.15.6.

2.28 Irreducible maps and the Auslander-Reiten Quiver

In this section, we describe a certain directed graph associated with the almost split sequences, and describe its elementary properties. The results of sections 2.28 to 2.32 are summarized in theorem 2.32.6.

2.28.1 Definition

Suppose U and V are indecomposable kG-modules. A map $\lambda: U \to V$ is irreducible if λ is not an isomorphism, and whenever $\lambda = \mu\nu$ is a factorization of λ, either μ has a left inverse or ν has a right inverse.

Let $\mathrm{Rad}(U,V)$ be the space of non-isomorphisms from U to V, and $\mathrm{Rad}^2(U,V)$ be the space spanned by the homomorphisms of the form $\alpha\beta$ with $\alpha \in \mathrm{Rad}(U,W)$ and $\beta \in \mathrm{Rad}(W,V)$ for some indecomposable W. Then the set of irreducible maps is precisely $\mathrm{Rad}(U,V) \setminus \mathrm{Rad}^2(U,V)$. The space $\mathrm{Irr}(U,V) = \mathrm{Rad}(U,V)/\mathrm{Rad}^2(U,V)$ is an $\mathrm{End}_{kG}(U)$ - $\mathrm{End}_{kG}(V)$ bimodule. Let its length as a left $\mathrm{End}_{kG}(U)$-module be a_{UV} and its length as a right $\mathrm{End}_{kG}(V)$-module be a'_{UV}. Note that if k is algebraically closed then

$a_{UV} = a'_{UV} = \dim \operatorname{Irr}(U,V)$.

The <u>Auslander-Reiten quiver</u> of G is the directed graph whose vertices are the indecomposable modules, and whose edges are as follows.

$$U \cdot \qquad \cdot V \qquad \text{if} \quad \operatorname{Irr}(U,V) = 0$$

$$U \xrightarrow{(a_{UV},\, a'_{UV})} V \qquad \text{if} \quad \operatorname{Irr}(U,V) \neq 0$$

$$U \longrightarrow V \qquad \text{if} \quad a_{UV} = a'_{UV} = 1$$

2.28.2 <u>Lemma</u>

If $\lambda: U \to V$ is irreducible, then λ is either an epimorphism with an indecomposable kernel, or a monomorphism with an indecomposable cokernel.

<u>Proof</u>

The factorization $U \xrightarrow{\mu} U/\operatorname{Ker}(\lambda) \xrightarrow{\nu} V$ shows that λ is either an epimorphism or a monomorphism. Suppose λ is an epimorphism with kernel $A \oplus B$. Then there is a factorization

$$U \xrightarrow{\mu} U/A \xrightarrow{\nu} U/(A \oplus B) = V.$$

If either μ has a left inverse or ν has a right inverse, it is easy to see that U is decomposable. A similar argument works when λ is a monomorphism. □

2.28.3 <u>Proposition</u>

(i) If V is not projective, let the almost split sequence terminating in V be $0 \longrightarrow \Omega^2 V \longrightarrow X_V \xrightarrow{\sigma} V \longrightarrow 0$. Then $\lambda: U \to V$ is irreducible if and only if U is a summand of X_V and $\lambda = i_U\, \sigma$ with $i_U: U \to X_V$ the inclusion.

(ii) If U is not projective, let the almost split sequence beginning with U be $0 \longrightarrow U \xrightarrow{\sigma'} X_{\mho^2 U} \longrightarrow \mho^2 U \longrightarrow 0$. Then $\lambda: U \to V$ is irreducible if and only if V is a direct summand of $X_{\mho^2 U}$ and $\lambda = \sigma' \pi_V$ with $\pi_V: X_{\mho^2 U} \to V$ the inclusion.

(iii) If U and V are both projective then there are no irreducible maps $U \to V$.

<u>Proof</u>

(i)

$$\begin{array}{ccccccccc} & & & & & & U & & \\ & & & & & \overset{\mu}{\swarrow} & \downarrow \lambda & & \\ 0 & \longrightarrow & \Omega^2 V & \longrightarrow & X_V & \xrightarrow{\sigma} & V & \longrightarrow & 0 \end{array}$$

Since λ is not an isomorphism, λ factors as $\mu\sigma$. Since σ does not have a right inverse, μ has a left inverse. Thus we may take $\mu = i_U$. Conversely suppose U is a direct summand of X_V with inclusion i_U, and $i_U\sigma = \mu\nu$.

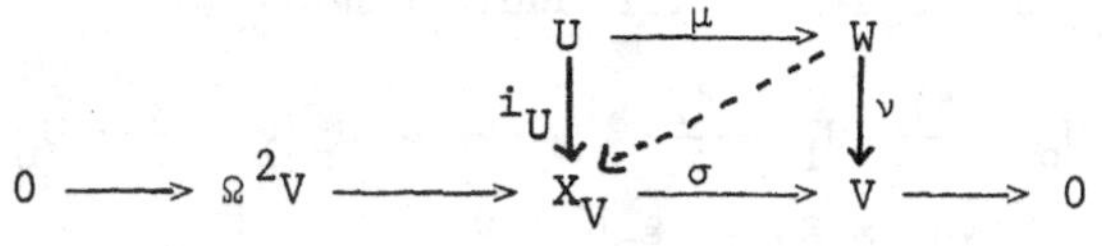

If ν does not have a right inverse, then ν factors through σ, and so μ has a left inverse.

(ii) This follows in a similar way from 2.17.8.

(iii) This follows from 2.28.2 and the fact that projective modules are injective. □

This proposition implies that the Auslander-Reiten quiver is a locally finite graph.

<u>Remark</u>

It follows from 2.28.3 that a_{UV} is the number of copies of V as a direct summand of the middle term of the almost split sequence starting with U, and that a'_{UV} is the number of copies of U as a direct summand of the middle term of the almost split sequence terminating with V.

2.28.4 <u>Lemma</u>

Suppose U and V are indecomposable kG-modules and $f: U \to V$ is not an isomorphism, and is non-zero.

(i) There is an irreducible map $g: U \to U'$ and a map $h: U' \to V$ with $gh \neq 0$.

(ii) There is a map $g: U \to V'$ and an irreducible map $h: V' \to V$ with $gh \neq 0$.

<u>Proof</u>

We shall prove (ii); (i) is proved dually using 2.17.8. Suppose V is not projective.

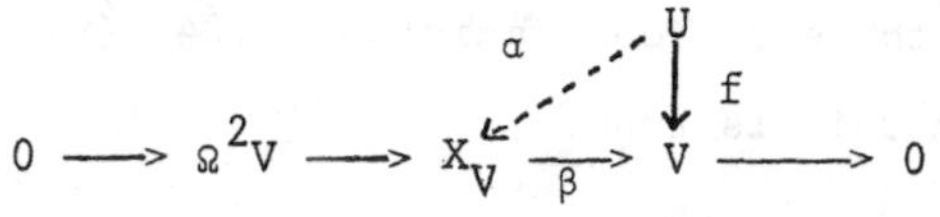

Write $X_V = \oplus X_i$ and $f = \alpha\beta = \Sigma\alpha_i\beta_i$ with $\alpha_i: U \to X_i$ and $\beta_i: X_i \to V$. Since $f \neq 0$, some $\alpha_i\beta_i \neq 0$, and β_i is an irreducible map. On the other hand if V is projective then f factors through the injection $\text{Rad}\, V \rightarrowtail V$. □

2.28.5 Proposition

Suppose U and V are indecomposable kG-modules, and $f:U \to V$ is not an isomorphism, and is non-zero. Suppose there is no chain of irreducible maps from U to V of length less than n.

(i) There exists a chain of irreducible maps

$$U = U_o \xrightarrow{g_1} U_1 \xrightarrow{g_2} \cdots \longrightarrow U_{n-1} \xrightarrow{g_n} U_n$$

and a map $h:U_n \to V$ with $g_1 g_2 \cdots g_n h \neq 0$.

(ii) There exists a chain of irreducible maps

$$V_n \xrightarrow{h_n} V_{n-1} \xrightarrow{h_{n-1}} \cdots \longrightarrow V_1 \xrightarrow{h_1} V_o = V$$

and a map $g: U \to V_n$ with $g h_n h_{n-1} \cdots h_1 \neq 0$.

Proof

This follows from 2.28.4 and induction. □

The projective modules often only get in the way when we are looking at the Auslander-Reiten quiver, and so we define the stable quiver to be the subgraph of the Auslander-Reiten quiver obtained by deleting the vertices corresponding to the projective modules and all edges meeting them.

Note that the only irreducible maps involving a projective indecomposable module P are $P \twoheadrightarrow P/\mathrm{Soc}(P)$ and $\mathrm{Rad}(P) \rightarrowtail P$. Thus if an almost split sequence involves a projective module, it is of the form $0 \to \mathrm{Rad}(P) \to P \oplus \mathrm{Rad}(P)/\mathrm{Soc}(P) \to P/\mathrm{Soc}(P) \to 0$.

2.28.6 Proposition

Any two modules in the same connected component of the stable quiver have the same complexity.

Proof

Clearly $cx_G(\Omega^2 V) = cx_G(V) = cx_G(\mho^2 V)$. Take V of minimal complexity in a connected component of the stable quiver, and adjacent to a module of strictly larger complexity. Then by 2.24.4(vi) and 2.28.3, we obtain an almost split sequence whose middle term has larger complexity than the ends, contradicting 2.24.4(v). □

In fact, more than this is true.

2.28.7 Proposition

If U and V are in the same connected component of the stable quiver, then $X_G(U) = X_G(V)$.

Proof

By 2.26.10 and 2.25.7, we only need check that for each cyclic subgroup P of kG of order p, $U{\downarrow}_P$ is free if and only if $V{\downarrow}_P$ is free. If this is false, then without loss of generality, there is a directed edge from U to V in the stable quiver. Suppose $V{\downarrow}_P$ is free while $U{\downarrow}_P$ is not. Then $(\Omega^2 V){\downarrow}_P$ is also free, and so $U{\downarrow}_P$ is a direct summand of an extension of a free module by a free module, and is hence free. Similarly if $U{\downarrow}_P$ is free while $V{\downarrow}_P$ is not, then $(\mho^2 U){\downarrow}_P$ is also free, and so $V{\downarrow}_P$ is again a direct summand of an extension of a free module by a free module. □

2.28.8 Corollary

Given a subset $X \subseteq \overline{X}_G$, the bilinear forms $(\, , \,)$ and $< \, , \, >$ are non-singular on $A(G,X)$.

Proof

If V is an indecomposable module with $X_G(V) \subseteq X$, then by 2.18.3, $v.\tau(V) \in A(G,X)$, and by 2.26.9(ii), $\tau(V) \in A(G,X)$. The result now follows from 2.18.4 as in the proof of 2.18.5. □

2.29 The Riedtmann Structure Theorem

We now wish to describe the structure theorem of Riedtmann (2.29.6). This theorem describes the structure of an abstract stable representation quiver, of which the stable quivers described in 2.28 are an example. The necessary terminology is given in the following definitions. The proof of the structure theorem involves a variant of the classical 'universal cover' construction.

2.29.1 Definitions

A quiver Q consists of a set of vertices Q_o, a set of arrows Q_1, and a pair of maps $d_o, d_1 \colon Q_1 \to Q_o$. For $\alpha \in Q_1$, we call $\alpha d_o \in Q_o$ the head of α and αd_1 the tail. For $x \in Q_o$, we set

$$x^+ = \{\alpha d_o : \alpha \in Q_1 \text{ and } \alpha d_1 = x\}$$

$$x^- = \{\alpha d_1 : \alpha \in Q_1 \text{ and } \alpha d_o = x\} .$$

A morphism of quivers $\phi \colon Q \to Q'$ is a pair of maps $\phi_o \colon Q_o \to Q'_o$ and $\phi_1 \colon Q_1 \to Q'_1$ such that the following squares commute.

$$\begin{array}{ccc} Q_1 & \xrightarrow{\phi_1} & Q_1' \\ {\scriptstyle d_o}\downarrow & & \downarrow{\scriptstyle d_o'} \\ Q_o & \xrightarrow{\phi_o} & Q_o' \end{array} \qquad\qquad \begin{array}{ccc} Q_1 & \xrightarrow{\phi_1} & Q_1' \\ {\scriptstyle d_1}\downarrow & & \downarrow{\scriptstyle d_1'} \\ Q_o & \xrightarrow{\phi_o} & Q_o' \end{array}$$

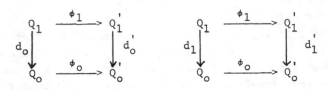

Q is called *locally finite* if x^+ and x^- are finite sets for all $x \in Q_o$.

To a quiver Q without *loops* or *double arrows* (i.e. subquivers of the form $x\circlearrowleft$ resp. $x \rightrightarrows y$) we associate a graph $\overline{Q}$ whose vertices are the vertices of Q, and two vertices x and y are joined by an edge if there is an arrow $x \to y$ or $x \leftarrow y$ in Q. A quiver is called a *directed tree* if Q has no subquiver of the form $x \rightleftarrows y$ and $\overline{Q}$ is a (connected) tree.

A *stable representation quiver* is a quiver Q together with an automorphism $\lambda: Q \to Q$ called the *translation* such that the following conditions are satisfied.

(i) Q contains no loops or double arrows.

(ii) For all $x \in Q_o$, $x^- = (x\lambda)^+$.

Example

The stable quiver of kG is a locally finite stable representation quiver, with translation Ω^2.

A *morphism* of stable representation quivers is a morphism of quivers commuting with the translation. A stable representation quiver is said to be *connected* if it is non-empty and cannot be written as a disjoint union of two subquivers each stable under translation (note that this does not imply that the underlying quiver is connected).

To a directed tree B we associate a stable representation quiver $\mathbb{Z}B$ as follows. The vertices of $\mathbb{Z}B$ are the pairs (n,x), $n \in \mathbb{Z}$, $x \in B_o$. For each arrow $x \to y$ and each $n \in \mathbb{Z}$, we have two arrows $(n,x) \to (n,y)$ and $(n,y) \to (n-1,x)$. The translation is defined via $(n,x)\lambda = (n+1,x)$. We regard B as embedded in $\mathbb{Z}B$ as $\{(0,x)\}$.

Examples

If B =  then $\overline{B}$ = 

and $\mathbb{Z}B$ = 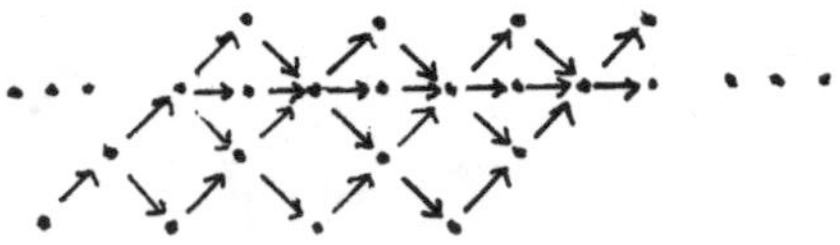

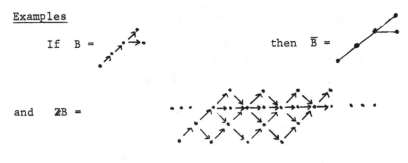

If $B =$ 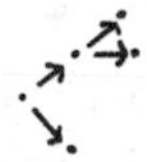 then $\overline{B}$ and $\mathbb{Z}B$ are again as above.

Keep this example in mind when reading the proofs of 2.29.3 and 2.29.6.

2.29.2 Lemma

Let B be a directed tree and Q a stable representation quiver. Given a quiver morphism $\phi: B \to Q$ there is a unique morphism of stable representation quivers $f: \mathbb{Z}B \to Q$ such that $(0,x)f = x\,\phi$.

Proof

$(n,x)f = x\phi\lambda^n$ is clearly the unique such morphism. □

2.29.3 Proposition

Let B and B' be directed trees. Then $\mathbb{Z}B \cong \mathbb{Z}B'$ if and only if $\overline{B} \cong \overline{B}'$.

Proof

$\overline{B}$ may be obtained from $\mathbb{Z}B/\lambda$ by replacing each double edge  by an undirected edge , and so $\mathbb{Z}B \cong \mathbb{Z}B'$ implies $\overline{B} \cong \overline{B}'$. Conversely suppose $\overline{B} \cong \overline{B}'$. Choose a point $x \in B_o$, and send it to $(0,x)$ in $\mathbb{Z}B_o'$. Since B is connected we may extend this uniquely to a morphism $B \to \mathbb{Z}B'$ in such a way that the induced morphism $\overline{B} \to \overline{B}'$ is our given isomorphism. Now by 2.29.2, we get a morphism of stable representation quivers $\mathbb{Z}B \to \mathbb{Z}B'$ which is clearly an isomorphism, since it sends (n,x) to $(n+a_x,x)$, where vertices of B and B' have been identified by the given tree isomorphism, and a_x are integer constants depending only on x. □

2.29.4 Definitions

A group Π of automorphisms of a stable representation quiver Q is called admissible if the orbit of a point x does not intersect $x^+ \cup x^-$. The quotient quiver Q/Π, defined in the obvious way, is clearly a stable representation quiver.

A morphism of representation quivers $f: Q \to Q'$ is called a covering if for each vertex $x \in Q_o$ the induced maps $x^- \to (xf)^-$ and $x^+ \to (xf)^+$ are bijective. It is clearly enough to check that $x^+ \to (xf)^+$ is bijective for each $x \in Q_o$.

Example

The canonical projection $Q \to Q/\Pi$, for Π an admissible group of automorphisms of Q, is a covering.

2.29.5 Lemma

Let B be a directed tree, $\pi : \mathbb{Z}B \to Q'$ a morphism of stable representation quivers, $\phi: Q \to Q'$ a covering, and (n,x) a vertex of $\mathbb{Z}B$. Then for each $y \in Q_o$ with $y\phi = (n,x)$, there is a unique morphism $\psi: \mathbb{Z}B \to Q$ with $\psi\phi = \pi$ and $y\phi = (n,x)\pi$.

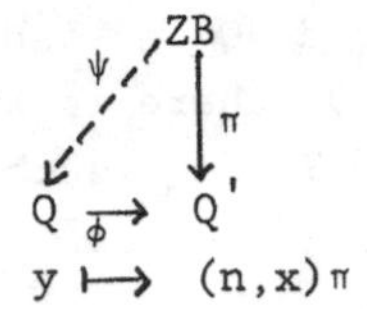

Proof

Renumber $\mathbb{Z}B$ so that $n = 0$. Then the map $(0,x)\psi = y$ clearly extends uniquely to a map from B to Q whose composite with ϕ is π. The result now follows from 2.29.2. □

2.29.6 Structure Theorem (Riedtmann)

For each connected stable representation quiver Q there is a directed tree B and an admissible group of automorphisms $\Pi \subseteq \mathrm{Aut}(\mathbb{Z}B)$ such that $Q \cong \mathbb{Z}B/\Pi$. The graph $\bar{B}$ associated to B is defined by Q uniquely up to isomorphism, and Π is uniquely defined up to conjugation in $\mathrm{Aut}(\mathbb{Z}B)$.

Proof

Given Q, we construct B as follows. Choose a point $x \in Q_o$, and let B have as vertices the paths

$$x \xrightarrow{\alpha_1} y_1 \xrightarrow{\alpha_2} y_2 \longrightarrow \dots \xrightarrow{\alpha_n} y_n \qquad (n \geq 0)$$

for which $y_i \neq y_{i+2}\lambda$ for $1 \leq i \leq n-2$. The arrows of B are

$$(x \xrightarrow{\alpha_1} y_1 \longrightarrow .. \xrightarrow{\alpha_{n-1}} y_{n-1}) \to (x \xrightarrow{\alpha_1} y_1 \longrightarrow .. \xrightarrow{\alpha_{n-1}} y_{n-1} \xrightarrow{\alpha_n} y_n).$$

Then clearly B is a directed tree.

The quiver morphism $B \to Q$ given by

$$(x \xrightarrow{\alpha_1} y_1 \longrightarrow .. \xrightarrow{\alpha_n} y_n) \longmapsto y_n$$

extends uniquely, by 2.29.2, to a morphism $\phi: \mathbb{Z}B \to Q$.

We check that $\phi : \mathbb{Z}B \to Q$ is a covering morphism. If $u = (x \xrightarrow{\alpha_1} y_1 \longrightarrow .. \xrightarrow{\alpha_n} y_n) \in B_o$ then

$$u^+ = \begin{cases} \{(x \xrightarrow{\alpha_1} y_1 \longrightarrow \cdots \xrightarrow{\alpha_n} y_n \xrightarrow{\beta} z) \\ \quad \text{such that} \quad z\lambda \neq y_{n-1}\} & n > 0 \\ \{(x \xrightarrow{\beta} z)\} & n = 0, \end{cases}$$

and so

$$(0,u)^+ = \{(0,v), v \in u^+\} \cup \{(-1,v), v \in u^-\}$$

has image $\{z \in y_n^+ : z\lambda \neq y_{n-1}\} \cup \{y_{n-1}\lambda^{-1}\} = y_n^+$ in Q. Hence $(m,u)^+$ has image $(y_n\lambda^m)^+$ as desired.

Now let Π be the <u>fundamental group</u> of Q at x, namely the group of morphisms of stable representation quivers $\rho: \mathbb{Z}B \to \mathbb{Z}B$ with $\rho\phi = \phi$. Since ϕ is a covering morphism, Π is admissible. Hence by 2.29.5, $Q \cong \mathbb{Z}B/\Pi$.

Also by 2.29.5, if $\mathbb{Z}B \to Q$ and $\mathbb{Z}B' \to Q$ are two such covers then we obtain inverse isomorphisms $\mathbb{Z}B \underset{g^{-1}}{\overset{g}{\rightleftarrows}} \mathbb{Z}B'$. Hence $\Pi' = g^{-1}\Pi g$, and by 2.29.3, $\bar{B} \cong \bar{B}'$. □

The stable representation quiver $\mathbb{Z}B$ is called the <u>universal cover</u> of Q, and the isomorphism class of B is called the <u>tree class</u> of Q.

We also have another graph associated with Q. We define the <u>reduced graph</u> Δ of Q to be the graph obtained from Q/λ by replacing each double edge  by an undirected edge . It is clear that an automorphism of Q determines an automorphism of Δ.

2.29.7 <u>Lemma</u>

There is a natural map $\varkappa$ from the tree $\bar{B}$ to the reduced graph Δ which satisfies

(i) $\varkappa$ is surjective

(ii) If x and y are adjacent points in $\bar{B}$ then $x\varkappa \neq y\varkappa$.

<u>Proof</u>

The composite map $\mathbb{Z}B \to Q \to \Delta$ is surjective, and has the property that (n,x) and $(n+1,x)$ have the same image. Thus it determines a well defined surjective map $\varkappa: \bar{B} \to \Delta$. Now $Q \cong B/\Pi$ with Π an admissible group of automorphisms, by 2.29.6, and so property (ii) follows from the definition of admissibility. □

We shall see in the next two sections that the tree class and reduced graph of a connected component of the stable quiver of kG are quite restricted in possible shape.

Now the translation Ω^2 on a component Q of the stable quiver of kG preserves the labelling (a_{UV}, a'_{UV}) on the edges, and so we have a labelling on Q/Ω^2. Moreover, since $a_{UV} = a'_{\Omega^2 V,U}$ and $a'_{UV} = a_{\Omega^2 V,U}$, we have a labelling (a_{ij}, a_{ji}) on the reduced graph. Lifting this back via $\varkappa$ gives a labelling on the associated tree.

2.30 Dynkin and Euclidean diagrams

In the last section, we saw that associated with each component of the stable quiver of kG we have a tree together with a labelling of the edges with pairs of numbers (a_{ij}, a_{ji}). In this section, we define the notion of a subadditive function on a labelled graph, and show that the existence of such a subadditive function imposes severe restrictions on the possible shape of the graph (theorem 2.30.6). In the next section, we shall construct a subadditive function on the labelled tree associated with a component of the stable quiver of kG, and then investigate the various possibilities given by theorem 2.29.6.

A labelled graph is a graph together with a pair of positive integers (a_{ij}, a_{ji}) for each edge i — j. As usual, we omit the labels when $a_{ij} = a_{ji} = 1$. We also use $\bullet\!\Rightarrow\!\bullet$ and $\bullet\!\Rrightarrow\!\bullet$ to signify the labelled edges $\bullet\overset{(2,1)}{\text{——}}\bullet$ and $\bullet\overset{(3,1)}{\text{——}}\bullet$ respectively. If T is a labelled graph then T^{op} is the labelled graph with $a^{op}_{ij} = a_{ji}$. A labelled tree is a labelled graph which is a tree. The Cartan matrix C_T of a labelled graph T (not to be confused with the Cartan matrix of an algebra) is the matrix whose rows and columns are indexed by the vertices of the graph, and with entries

$$c_{ij} = \begin{cases} 2 & \text{if } i = j \\ -a_{ij} & \text{if } i \text{ —— } j \text{ is an edge} \\ 0 & \text{otherwise} \end{cases} .$$

We shall be interested in the following labelled graphs.

(i) The finite Dynkin diagrams

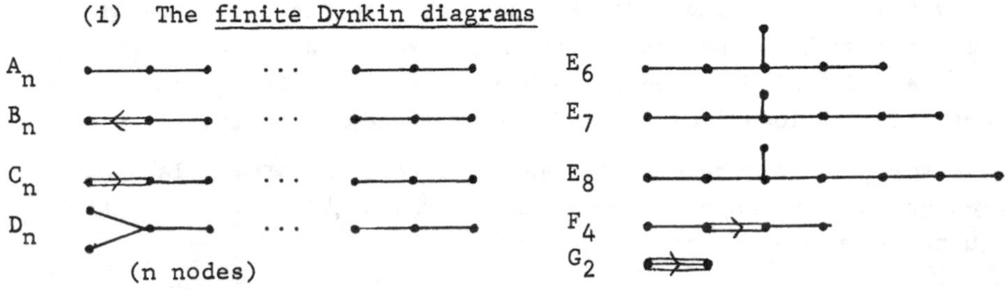

(ii) The <u>infinite Dynkin diagrams</u>

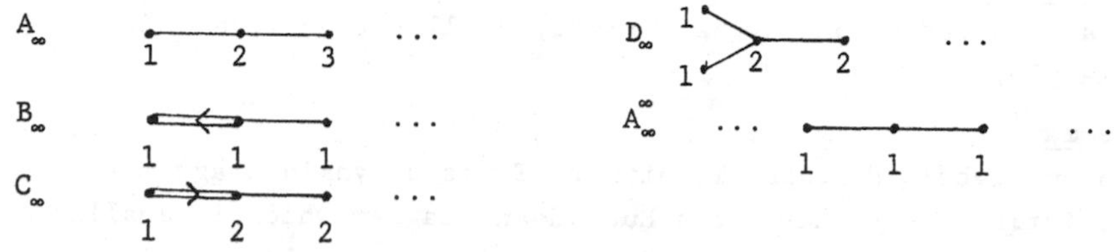

(iii) The <u>Euclidean diagrams</u>

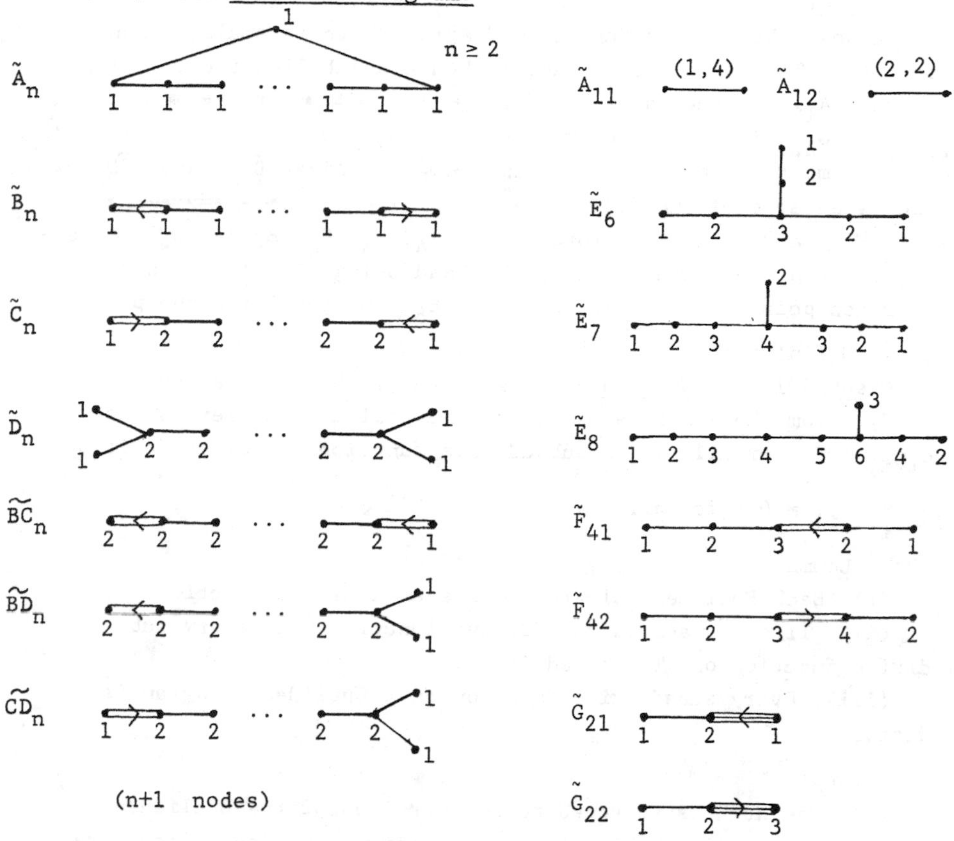

<u>Remark</u>

For the moment, ignore the numbers attached to the vertices of the Euclidean and infinite Dynkin diagrams. These will appear in the proofs of 2.30.3 and 2.30.5.

Given two labelled graphs T_1 and T_2, we say that T_1 is <u>smaller</u> than T_2 if there is an injective morphism of graphs

$\rho: T_1 \to T_2$ such that for each edge i —— j in T_1, $a_{ij} \le a_{i\rho,j\rho}$, and <u>strictly smaller</u> if ρ can be chosen not to be an isomorphism. Note that a labelled graph may be strictly smaller than itself (see A_∞ for example).

2.30.1 <u>Lemma</u>

Given any labelled graph T, either T is a Dynkin diagram (finite or infinite) or there is a Euclidean diagram which is smaller than T (both possibilities may not occur simultaneously).

<u>Proof</u>

Suppose there is no Euclidean diagram which is smaller than T. Using $\tilde{A}_n$, T has no cycles, and is hence a labelled tree. Using $\tilde{A}_{11}$ and $\tilde{A}_{12}$ all edges of the form $\bullet\!\!-\!\!\bullet$, $\bullet\!\!\Rightarrow\!\!\bullet$ or $\bullet\!\!\Rrightarrow\!\!\bullet$. Using $\tilde{G}_{21}$ and $\tilde{G}_{22}$, if $\bullet\!\!\Rrightarrow\!\!\bullet$ occurs then $T \cong G_2$. Using $\tilde{B}_n$, $\tilde{C}_n$ and $\widetilde{BC}_n$, T has at most one edge of the form $\bullet\!\!\Rightarrow\!\!\bullet$. Using $\widetilde{BD}_n$ and $\widetilde{CD}_n$, if there is an edge of the form $\bullet\!\!\Rightarrow\!\!\bullet$ then $T = \ldots$ $\bullet\!\!-\!\!\bullet\!\!\Rightarrow\!\!\bullet\!\!-\!\!\bullet$ $\ldots$; using $\tilde{F}_{41}$ and $\tilde{F}_{42}$ this forces $T \cong F_4$, B_n, C_n, B_∞ or C_∞. Otherwise T is a tree with single edges, and using $\tilde{D}_n$, it has at most one branch point. Using $\tilde{E}_6$, $\tilde{E}_7$ and $\tilde{E}_8$ now completes the proof. □

2.30.2 <u>Definition</u>

A <u>subadditive function</u> on a labelled graph T is a function $t_i \mapsto d_i$ from the vertices of T to the positive integers satisfying $\sum_i d_i c_{ij} \ge 0$ for all j. A subadditive function is called <u>additive</u> if $\sum_i d_i c_{ij} = 0$ for all j.

2.30.3 <u>Lemma</u>

(i) Each Euclidean diagram admits an additive function.

(ii) If T^{op} admits an additive function then every subadditive function on T is additive.

(iii) Every subadditive function on a Euclidean diagram is additive.

<u>Proof</u>

(i) The numbers attached to the vertices of the Euclidean diagrams in the illustration form an additive function in each case.

(ii) Suppose d is a subadditive function on T. By hypothesis there is a function d' such that $\sum_j c_{ij} d'_j = 0$ for all i. Thus

$0 = \sum_{i,j} d_i c_{ij} d'_j$, while $\sum_i d_i c_{ij} \ge 0$ and $d'_j > 0$ for each j.

Hence d is additive.

(iii) follows from (i) and (ii). □

2.30.4 Lemma

Suppose T and T' are connected labelled graphs and T is strictly smaller than T'. Suppose also that d is a subadditive function on T'. Then identifying T with a subgraph of T', d restricted to T is a subadditive function on T which is not additive.

Proof

For j a vertex of T,

$$2d_j \geq \sum_{\substack{i \in T' \\ i \neq j}} d_i a_{ij}^{(T')} \geq \sum_{\substack{i \in T \\ i \neq j}} d_i a_{ij}^{(T)} \quad .$$

Since T is strictly smaller than T', for some $j \in T$ the right hand inequality is strict, and so the restriction of d is not additive. □

2.30.5 Lemma

Each of the infinite Dynkin diagrams admits an additive function.

(i) For A_∞ there are also subadditive functions which are not additive.

(ii) For the other infinite Dynkin diagrams every subadditive function is a multiple of a given bounded additive function.

Proof

The numbers attached to the vertices in the illustration form an additive function in each case.

(i) A_∞ is strictly smaller than itself, and so by 2.30.4 there is a subadditive function which is not additive.

(ii) For A_∞^∞, given a subadditive function d, choose a vertex i with d_i minimal. Then the sum of the two adjacent d_j is at most $2d_i$, and so each equals d_i. Inductively we find that the function is constant and additive.

For B_∞, C_∞ and D_∞, given a subadditive function we generate a subadditive function on A_∞^∞ as follows.

$$d_o \Leftarrow d_1 — d_2 — \cdots \longmapsto \cdots — d_2 — d_1 — d_o — d_1 — d_2 — \cdots$$

$$d_o \Rightarrow d_1 — d_2 — \cdots \longmapsto \cdots — d_2 — d_1 — 2d_o — d_1 — d_2 — \cdots$$

$$\begin{array}{l} d_o \diagdown \\ \qquad\quad d_1 — d_2 — \cdots \longmapsto \cdots — d_2 — d_1 — d_o'+d_o' — d_1 — d_2 — \cdots \\ d_o' \diagup \end{array}$$

The result follows immediately for B_∞ and C_∞. For D_∞ we obtain $d_o + d_o' = d_1$. Subadditivity forces $2d_o \geq d_1$ and $2d_o' \geq d_1$ whence $d_o = d_o'$, and the result follows. □

The following is a generalization by Happel, Preiser and Ringel of Vinberg's characterization of the finite Dynkin and Euclidean diagrams. See also [15].

2.30.6 Theorem

Let T be a connected labelled graph, and d a subadditive function on T. Then

(i) T is either a Dynkin diagram (finite or infinite) or a Euclidean diagram.

(ii) If d is not additive then T is a finite Dynkin diagram or A_∞.

(iii) If d is additive then T is an infinite Dynkin diagram or a Euclidean diagram.

(iv) If d is unbounded then $T \cong A_\infty$.

Proof

(i) Suppose false. Then by 2.30.1 there is a Euclidean diagram which is strictly smaller than T. Thus by 2.30.4 there is a subadditive function on this Euclidean diagram which is not additive, contradicting 2.30.3 (iii).

(ii) This follows from 2.30.3 (iii) and 2.30.5 (ii).

(iii) Suppose false. Then T is a finite Dynkin diagram by (i), and hence so is T^{op}. Thus T^{op} is strictly smaller than some Euclidean diagram. Thus by 2.30.3 (i) and 2.30.4 T^{op} admits a subadditive function which is not additive, contradicting 2.30.3 (ii).

(iv) If d is unbounded then T is infinite, and so by (i) it is an infinite Dynkin diagram. Hence by 2.30.5 (ii) $T \cong A_\infty$. □

2.31 The tree class of a connected component of the stable quiver

We now wish to construct a subadditive function on the labelled tree determined by a connected component of the stable quiver of kG, and to determine when this function is additive. In order to do this, we take another look at Poincaré series. In 2.22 we saw that $\xi_V(t)$ is a rational function of the form $f(t)/\prod_{i=1}^{r}(1-t^{k_i})$. We can also form another Poincaré series $\eta_V(t) = \Sigma\, t^n \dim(P_n)$, where $\dots \to P_2 \to P_1 \to P_o \to V \to 0$ is a minimal projective resolution of V. Since for S simple the number of times P_S occurs as a summand of P_n is $\dim_k \mathrm{Ext}^n_G(V,S)/\dim_k \mathrm{End}_{kG}(S)$, we have

$$\dim_k(P_n) = \sum_{\substack{S \\ \text{simple}}} \frac{\dim_k(P_S).\dim_k \mathrm{Ext}_G^n(V,S)}{\dim_k \mathrm{End}_{kG}(S)} .$$

Now, $\mathrm{Ext}_G^*(V,S) \cong H^*(G,V^* \otimes S)$ is a finitely generated module for $H^*(G,k)$ by Evens' theorem (see 2.22, properties of cohomology (vii)). Thus by 1.8.2, $\eta_V(t)$ is a rational function of the form $f(t)/\prod_{i=1}^{r}(1-t^{k_i})$, where the k_i are independent of V, and f has integer coefficients. It now follows from 1.8.3 that the pole of $\eta_V(t)$ at $t = 1$ has order $c = cx_G(V)$, and that the value of the analytic function $(\Pi k_i)\,\eta_V(t)(1-t)^c$ at $t = 1$ is a positive integer. We denote this value by $\eta(V)$. This will in fact form our subadditive function.

2.31.1 Proposition

(i) $\eta(V) = \eta(\Omega V)$

(ii) If $0 \to V' \to V \to V'' \to 0$ is a short exact sequnece of modules of the same complexity (cf. 2.24.4(v)), then

$$\eta(V) \le \eta(V') + \eta(V'').$$

(iii) If $0 \to \Omega^2 V \to X_V \to V \to 0$ is an almost split sequence, then $\eta(X_V) \le 2\eta(V)$; if $\eta(X_V) < 2\eta(V)$ then V is periodic, and for some n the middle term $X_{\Omega^n V}$ of the almost split sequence $0 \to \Omega^{n+2}V \to X_{\Omega^n V} \to \Omega^n V \to 0$ has a projective direct summand.

Proof

(i) $\eta_V(t) = t\,\eta_{\Omega V}(t) + \text{constant}$.

(ii) Without loss of generality V' and V'' have no projective direct summands (although V may have) since these split off from the sequence without affecting $\eta(V') + \eta(V'') - \eta(V)$. Thus if $.. \to P_1' \to P_o' \to V' \to 0$ and $.. \to P_1'' \to P_o'' \to V'' \to 0$ are minimal projective resolutions then there is a projective resolution $.. \to P_1' \oplus P_1'' \to P_o' \oplus P_o'' \to V \to 0$. Thus $\eta_{V'}(t) + \eta_{V''}(t) - \eta_V(t)$ is the Poincaré series of a graded module, and so the value at $t = 1$ of $(\eta_{V'}(t) + \eta_{V''}(t) - \eta_V(t))(1-t)^c$ cannot be negative (see 1.8.3).

(iii) By 2.28.6, V, X_V and $\Omega^2 V$ have the same complexity. $X_{\Omega^n V} \cong \Omega^n(X_V)$ unless and only unless $X_{\Omega^n V}$ has a projective direct summand. Thus the minimal projective resolution of X_V is the sum of the minimal projective resolutions of V and $\Omega^2 V$ except at those places where $X_{\Omega^n V}$ has a projective direct summand. If this happens for only finitely many values of n, then $\eta_V(t) + \eta_{\Omega^2 V}(t) - \eta_{X_V}(t)$ is a polynomial in t, and hence $\eta(X_V) = \eta(V) + \eta(\Omega^2 V) - 2\eta(V)$. Otherwise, V must be periodic, since there are only finitely many projective indecomposables, each of which appears in only one almost split sequence. □

2.31.2 Theorem

(i) (Webb, [92]) Let T be the tree class of a connected component Q of the stable quiver of kG. Then T is either a Dynkin diagram (finite or infinite) or a Euclidean diagram (apart from $\tilde{A}_n$).

(ii) The reduced graph Δ of Q is also either a Dynkin diagram (finite or infinite) or a Euclidean diagram (this time $\tilde{A}_n$ is allowed).

Proof

By 2.31.1, $\eta(V)$ defines a function on Q which commutes with Ω^2, and satisfies $2\eta(x) \geq \sum_{y \in x^-} \eta(y)$. Thus η gives a subadditive function on both T and Δ. The result thus follows from 2.30.6 (i). □

Remark

If k is algebraically closed then each $a_{UV} = a'_{UV}$ (see definition 2.28.1) since the irreducible modules for $\mathrm{End}_{kG}(U)$ and $\mathrm{End}_{kG}(V)$ are one dimensional. Thus only the diagrams of type A, D, E, $\tilde{A}$ (but not $\tilde{A}_{11}$), $\tilde{D}$ and $\tilde{E}$ occur.

2.31.3 Corollary

The length of $\mathrm{Irr}(U,V)$ as a left $\mathrm{End}_{kG}(U)$-module and as a right $\mathrm{End}_{kG}(V)$-module are at most four, and if k is algebraically closed, $\dim_k \mathrm{Irr}(U,V) \leq 2$.

Proof

Among the list of Dynkin and Euclidean diagrams, the maximum a_{ij} appearing is four, and if $a_{ij} = a_{ji}$ then the maximum value appearing is two; indeed, we may further observe that $a_{ij} \cdot a_{ji}$ is at most four. □

2.31.4 Corollary

The number of direct summands in the middle term of an almost split sequence is at most five, and if equal to five, then one of the summands is projective.

Proof

Among the Dynkin and Euclidean diagrams, the maximal possible value of $\sum_j a_{ij}$ is four. □

2.31.5 Corollary

Let P be a (non-simple) projective indecomposable kG-module. Then the maximal possible number of direct summands of Rad(P)/Soc(P) is four.

Proof

Apply 2.31.4 to the almost split sequence

$$0 \to \text{Rad}(P) \to P \oplus \text{Rad}(P)/\text{Soc}(P) \to P/\text{Soc}(P) \to 0. \qquad \square$$

Following Webb, we now investigate each of the possibilities allowed by 2.31.2 in turn.

Case 1 The Finite Dynkin Diagrams

2.31.6 Lemma

Suppose a connected component Q of the stable quiver of kG contains a periodic module. Then every module in Q is periodic.

Proof

This follows from 2.25.4 and 2.24.4(xi). Alternatively, we may prove this directly as follows. If x is periodic with $x = \Omega^{2n}x$ then Ω^{2n} induces a permutation on x^-, which is a finite set by 2.28.3, and so some power of Ω^{2n} stabilizes x^- pointwise. Hence by induction all modules in Q are periodic. □

2.31.7 Proposition

Suppose the tree class T or the reduced graph Δ of a component Q of the stable quiver of kG is a finite Dynkin diagram. Then Q has only finitely many vertices.

Proof

Suppose T or Δ is a finite Dynkin diagram. Then by 2.30.6 η defines a subadditive function on T or Δ which is not additive. Hence for some module V belonging to Q, $\eta(X_V) < 2\eta(V)$. Thus by 2.31.1(iii), V is periodic. Hence by 2.31.2 every module in Q is periodic, and so Q has only finitely many vertices. □

2.31.8 Proposition

Suppose a component Q of the stable quiver of kG has only finitely many vertices. Then Q consists of all the non-projective modules in a block of kG with cyclic defect group.

Proof

Let A be the linear span in A(G) of the modules in Q. Suppose the modules in A do not constitute the set of non-projective modules in a complete block. Then there is a non-projective indecomposable module V outside A and a module W in A such that there is a non-zero homomorphism from V to W. Now by 2.18.3, for each indecomposable module U in A, $v.\tau(U) \in A$, and so by 2.18.4 (,) is non-singular on A. Since A is finite dimensional, this means that there exists $x \in A$ such that $(x,U) = (V,U)$ for all $U \in A$. Let W_o be an indecomposable module in A such that x has a non-zero coefficient of W_o. Then

$$0 \neq (x, v.\tau(W_o)) \qquad \text{(see 2.18.5)}$$
$$= (V, v.\tau(W_o)).$$

Thus by 2.18.4 $V = W_o$ since V and W_o are indecomposable. This contradiction shows that Q consists of all the non-projective modules in a block B of kG. The fact that B has cyclic defect group follows from 2.12.9. □

Remark

In fact in [77] and [52] it is shown that if B is a block of kG with cyclic defect group then the tree class of the corresponding connected component of the stable quiver is equal to the reduced graph, and is the Dynkin diagram A_n. However, the other finite Dynkin diagrams come up in algebras of finite representation type which are not blocks of finite group algebras.

Case 2 Infinite Quiver Components

In case 1, we saw that either $T \cong A_n$ or Q has infinitely many vertices. We shall now show that if Q is infinite then there are indecomposable modules in Q with an arbitrarily large number of composition factors. We shall then give two applications of this. First, we shall show that in the special case where Q has a periodic module, $T \cong A_\infty$. We shall then go on to consider the Euclidean diagrams, and show that if T is Euclidean then there is a projective module attached to Q.

2.31.9 Lemma (Harada, Sai)

Let $V_o, \ldots, V_{2^n-1}$ be indecomposable modules, each having at most n composition factors, and suppose $f_i : V_{i-1} \to V_i$ is not an isomorphism. Then $f_1 f_2 \ldots f_{2^n-1} = 0$.

Proof

Write $|U|$ for the number of composition factors of a module U (and $|0| = 0$). We show by induction on m that $|\mathrm{Im}(f_1 \ldots f_{2^m-1})| \leq n-m$. The assertion is clear for $m = 1$, since f_1

is not an isomorphism. Suppose true for $m - 1$. Write $f = f_1 \ldots f_{2^{m-1}-1}$, $g = f_{2^{m-1}}$ and $h = f_{2^{m-1}+1} \ldots f_{2^m-1}$. By the inductive hypothesis $|\mathrm{Im}(f)| \le n - m + 1$ and $|\mathrm{Im}(h)| \le n - m + 1$. If either inequality is strict, we are done, so suppose they are both equalities, and suppose $|\mathrm{Im}(fgh)| = n - m + 1$. Then $\mathrm{Im}(f) \cap \mathrm{Ker}(gh) = 0$ and $\mathrm{Im}(fg) \cap \mathrm{Ker}(h) = 0$. Thus by counting composition factors, $V_{2^m-1} = \mathrm{Im}(f) \oplus \mathrm{Ker}(gh)$ and $V_{2^m} = \mathrm{Im}(fg) \oplus \mathrm{Ker}(h)$. Since each is indecomposable, gh is injective and fg is surjective. Thus g is an isomorphism, contrary to hypothesis. □

2.31.10 <u>Theorem</u> (Auslander)

Suppose Q is an infinite component of the stable quiver of kG. Then Q has modules with an arbitrary large number of composition factors.

<u>Proof</u>

Suppose to the contrary that all modules in Q have at most n composition factors. Suppose U and V are indecomposable modules and $(U,V) \ne 0$ (recall $(U,V) = \dim_k \mathrm{Hom}_{kG}(U,V)$). If $U \in Q$, then also $V \in Q$. For if $V \notin Q$, then by 2.27.5 there is a chain of irreducible maps

$$U = U_0 \xrightarrow{g_1} U_1 \xrightarrow{g_2} \ldots \xrightarrow{g_{2^n-1}} U_{2^n-1}$$

and a map $h : U_{2^n-1} \to V$ with $g_1 \ldots g_{2^n-1} h \ne 0$, contradicting 2.31.9. The dual argument also shows that if $V \in Q$ then $U \in Q$.

Now for any indecomposable module V in Q, there is a projective module P with $(P,V) \ne 0$, and hence $P \in Q$. Thus every module is Q is connected by a chain of irreducible maps of length at most 2^n-1 to a projective module. Since there are only finitely many projectives in Q, and Q has finite valence, this shows that Q is finite, contrary to assumption. □

2.31.11 <u>Theorem</u>

Suppose an infinite component Q of the stable quiver of kG has a periodic module . Then the tree class of Q is A_∞.

<u>Proof</u>

By 2.31.6, every module in Q is periodic. If V is periodic, then $\eta(V)$ may be expressed in the form

$(\Pi k_i).$(average dimension of $\Omega^n(V)$).

Since Q is infinite, 2.31.10 shows that $\eta(V)$ is unbounded. Thus by 2.30.6(iv), the tree class of Q is A_∞. □

Finally, we examine the Euclidean diagrams in the next section.

2.32 Weyl groups and Coxeter transformations

In this section, we examine the geometry of a certain rational vector space associated with the graphs discussed in 2.30. Our goal is to prove proposition 2.32.4 and theorem 2.32.5. Finally, we summarize the results of sections 2.28 and 2.32 in our final theorem, 2.32.6.

A valued graph T is a finite labelled graph such that there exist natural numbers f_i, one for each vertex of T, with $a_{ij}f_j = a_{ji}f_i$, for each edge i —— j in T. Note that all the finite Dynkin and Euclidean diagrams are valued graphs, and that the numbers f_i, when they exist, are uniquely defined up to constant multiplication on each connected component. The matrix $\tilde{C}_T = (c_{ij}f_j)$ is called the symmetrized Cartan matrix, and is self-transpose.

Given a valued graph T, we form the rational vector space $\mathbb{Q}^T$ with the points t_i of T as basis, and we bestow $\mathbb{Q}^T$ with the symmetric bilinear form given by $\tilde{C}_T$:

$$<\underline{x},\underline{y}> = \sum_{i,j} c_{ij}f_jx_iy_j$$

($\underline{x} = \Sigma\, x_it_i$, $\underline{y} = \Sigma\, y_it_i$).

The Weyl group $W(T)$ is the group generated by the reflections

$$\underline{x}\, w_i = \underline{x} - 2\,\frac{<\underline{x},t_i>}{<t_i,t_i>}\, t_i .$$

It is easy to check that the w_i are transformations of order two preserving the bilinear form. By examining the two point graphs we see that the order of w_iw_j is 2, 3, 4, 6 or ∞ for $a_{ij}\, a_{ji} = 0,1,2,3$ or ≥ 4 respectively.

2.32.1 Lemma

Let T be a connected valued graph.

(i) T is a finite Dynkin diagram if and only if $<\ ,\ >$ is positive definite on $\mathbb{Q}^T$.

(ii) T is a Euclidean diagram if and only if $<\ ,\ >$ is positive semidefinite on $\mathbb{Q}^T$. In this case every null vector is a multiple of the vector given by the additive function shown in 2.30.

Proof

Suppose T is a Euclidean diagram. By 2.30.3(i), there is an additive function $t_i \mapsto d_i$ on T. Thus we have

$$< \underline{x},\underline{x} > = -\frac{1}{2} \sum_{i \neq j} d_i d_j c_{ij} f_j \left(\frac{x_i}{d_i} - \frac{x_j}{d_j} \right)^2$$

which is positive semidefinite since the c_{ij} are negative for $i \neq j$, and the d_i and f_i are positive. Moreover for a null vector, all the $\frac{x_i}{d_i}$ must have the same value so that the null space is one dimensional.

Since every Dynkin diagram is strictly smaller than a Euclidean diagram, it follows that $< , >$ is positive definite on the Dynkin diagrams.

If T is neither Euclidean nor Dynkin then by 2.30.1 there is a Euclidean diagram T' which is strictly smaller than T. If T' contains all the points of T, then a null vector for T' has negative norm for T. Otherwise choose a point of T adjacent to a point of T', and add a small enough multiple of the corresponding basis element to the null vector for T', to obtain a vector of negative norm. □

2.32.2 Lemma

Suppose T is a Euclidean diagram. Let $\underline{n}$ be the null vector given by the additive function shown in 2.30. Then $W(T)$ preserves $<\underline{n}>$ and acts as a finite group of automorphisms of $\mathbb{Q}^T/<\underline{n}>$.

Proof

Since $<\underline{n}>$ is the radical of $< , >$, $<\underline{n}>$ is preserved by $W(T)$. Since the matrices in $W(T)$ have integer entries with respect to our basis t_i, $W(T)$ acts as a discrete subgroup of the compact orthogonal group on $\mathbb{Q}^T/<\underline{n}>$, and this action is therefore finite. □

We now define a Coxeter transformation to be a product of all the w_i, taken once each in some order. Let T be a Euclidean diagram and let c be a Coxeter transformation. By 2.32.2, c has finite order m on $\mathbb{Q}^T/<\underline{n}>$, and so we may define the defect $\partial_c(\underline{x})$ of a vector $\underline{x} \in \mathbb{Q}^T$ via

$$\underline{x}\, c^m = \underline{x} + \partial_c(\underline{x})\, \underline{n}.$$

Thus ∂_c is a linear form $\mathbb{Q}^T \to \mathbb{Q}$ and $\partial_c(\underline{x}) = \sum_i \partial_c(t_i) x_i$.

The map ∂_c gives us a splitting $\mathbb{Q}^T = \mathrm{Ker}(\partial_c) \oplus <\underline{n}>$.

2.32.3 Lemma

The following two conditions on a vector $\underline{x} \in \mathbb{Q}^T$ are equivalent.

(i) $\underline{x}$ has infinitely many images under c

(ii) $\partial_c(\underline{x}) \neq 0$.

If (i) and (ii) are satisfied then some image of $\underline{x}$ has negative coordinates.

Proof

This is clear from the preceding discussion. □

Now let $\bar{B}$ be a directed labelled tree. A slice of $\mathbb{Z}B$ (see section 2.29) is a connected subgraph of $\mathbb{Z}B$ containing one representative of each point in B. If S is a slice, we write S^+ for the adjacent slice $\{(n+1,x) : (n,x) \in S\}$.

An additive function on $\mathbb{Z}B$ is a function f from the vertices of $\mathbb{Z}B$ to the positive integers with the property that

$$f(x) + f(x\lambda) = \sum_{y \in x^-} f(y).a_{yx}$$

where a_{yx} is to be interpreted as the number a_{ij} where t_i is the image of y and t_j is the image of x in $\bar{B}$.

2.32.4 Proposition

If $\bar{B}$ is a Euclidean tree then every additive function on $\mathbb{Z}B$ takes bounded values.

Proof

Let f be an additive function on $\mathbb{Z}B$. If S is a slice of $\mathbb{Z}B$ then we have a corresponding vector $\underline{x}_S \in \mathbb{Q}^{\bar{B}}$ whose i^{th} coordinate x_i is the value of f on the unique vertex in S lying above $t_i \in \bar{B}$. It is easy to check that if the vertex $(n,t_i) \in S$ is a sink (i.e. all directed edges in S involving (n,t_i) go towards (n,t_i)) then the slice $S.w_i$ obtained by replacing (n,t_i) by $(n+1,t_i)$ in S satisfies

$$\underline{x}_{S.w_i} = \underline{x}_S \, w_i$$

by the definition of additivity of f. Since $\bar{B}$ is a tree, we may choose an ordering for the vertices of $\bar{B}$ in such a way that each t_i is a sink for $S.w_1 \ldots w_{i-1}$. Thus the associated Coxeter element c takes S to the adjacent slice S^+. Since f only takes positive values it follows from 2.32.3 that $\underline{x}_S$ has only finitely many images under c. This implies that f takes on only finitely many different values on $\mathbb{Z}B$. □

2.32.5 Theorem

Suppose Q is a connected component of the stable quiver of kG, whose tree class is a Euclidean diagram. Then there is a projective module attached to Q.

Proof

Let $\mathbb{Z}B$ be the universal cover of Q, with $\overline{B}$ a Euclidean diagram. Suppose there is no projective module attached to Q. Then the dimension function on Q lifts to an additive function on $\mathbb{Z}B$. Thus by 2.32.4, the dimensions of modules in Q are bounded, contradicting 2.31.10 (Q has infinitely many vertices by case 1 of 2.31). □

Finally, the following theorem summarizes the results of sections 2.28 to 2.32.

2.32.6 Theorem

Let Q be a connected component of the stable quiver of kG. Then associated with Q we have a tree class T and a reduced graph Δ, both of which are labelled graphs, together with a natural surjective map $\varkappa: T \to \Delta$, which never identifies adjacent vertices of T. Each of T and Δ is either a Dynkin diagram (finite or infinite) or a Euclidean diagram.

(i) T is a Dynkin diagram if and only if Δ is a Dynkin diagram, which in turn happens if and only if the modules in Q belong to a block B with cyclic defect group. In this case Q consists of all the non-projective modules in B, $\varkappa: T \to \Delta$ is an isomorphism, and $T \cong A_n$.

(ii) If T is not a Dynkin diagram then there are indecomposable modules in Q of arbitrarily large dimension.

(iii) If Q contains a periodic module then every module in Q is periodic and $T \cong A_\infty$.

(iv) If T is a Euclidean diagram then there is a projective module attached to Q. In particular, only finitely many connected components of the stable quiver have a Euclidean diagram as their tree class. □

2.33 Galois Descent on the Stable Quiver

In 2.29, we introduced the concepts of tree class and reduced graph for a connected component of the stable quiver of kG-modules.

In this section we investigate what happens under field extensions, and we see that the reduced graph behaves much better than the tree class.

Let k be a field of characteristic p and let K be a finite Galois extension of k. Then $\mathfrak{G} = \mathrm{Gal}(K/k)$ acts on the set of indecomposable KG-modules as follows. If V is a KG-module and $\sigma \in \mathfrak{G}$, let V^{σ} be the representation with the same underlying set and the same action of G, but with new scalar multiplication by λ equal to the old scalar multiplication by λ^{σ}. It is clear that σ sends almost split sequences to almost split sequences and irreducible morphisms to irreducible morphisms. Thus $\mathfrak{G}$ acts as automorphisms on the stable quiver of KG-modules, and hence also permutes the connected components.

Denote by $e_{k,K}$ the natural map $A_k(G) \to A_K(G)$ given by $V \mapsto V \otimes_k K$, and by $f_{K,k}$ the natural map $A_K(G) \to A_k(G)$ given by $V \mapsto (1/|K:k|)V{\downarrow}_{kG}$.

2.33.1 <u>Lemma</u>

(i) $e_{k.K}$ and $f_{K,k}$ are ring homomorphisms.

(ii) $e_{k,K}$ is injective (see also exercise to 2.18).

(iii) $A_K(G) = \mathrm{Im}(e_{k,K}) \oplus \mathrm{Ker}(f_{K,k})$ as a direct sum of ideals.

(iv) $e_{k,K}$ preserves the inner products $(\; , \;)$ and $< \; , \; >$. Also, $(e_{k,K}x,y) = (x,f_{K,k}y)$ and $<e_{k,K}x,y> = <x,f_{K,k}y>$.

Because of (i) - (iv), we shall identify $A_k(G)$ with its image under $e_{k,K}$ from now on.

(v) If V is an indecomposable kG-module and $V \otimes_k K = V_1 \oplus .. \oplus V_n$ then $\mathfrak{G} = \mathrm{Gal}(K/k)$ acts transitively on the isomorphism types of the V_i.

(vi) $A_k(G)$ is the set of fixed points of $\mathfrak{G}$ on $A_K(G)$ (but note that $a_k(G)$ is in general smaller than the set of fixed points of $\mathfrak{G}$ on $a_K(G)$).

(vii) $e_{k,K}$ commutes with the map τ defined in section 2.18.

<u>Proof</u>

(i) It is clear that $e_{k,K}$ is a ring homomorphism. Since $V{\downarrow}_{kG} \otimes_k W{\downarrow}_{kG} \cong ((V{\downarrow}_{kG} \otimes_k K) \otimes_K W){\downarrow}_{kG} = |K:k|(V \otimes_K W){\downarrow}_{kG}$, $f_{K,k}$ is also a ring homomorphism.

(ii) and (iii) follow from the fact that $e_{k,K}$ followed by $f_{K,k}$ is the identity map.

(iv) This follows from the identities

$$\mathrm{Hom}_{kG}(V,W) \underset{k}{\otimes} K = \mathrm{Hom}_{KG}(V \underset{k}{\otimes} K,\ W \underset{k}{\otimes} K)$$

and $e_{k,K}(u_{kG}) = u_{KG}$, together with 2.4.3.

(v) Suppose $\mathfrak{G}$ does not act transitively on the isomorphism types of the V_i. Reorder the V_i so that for some k with $1 \le k \le n$, no one of $V_1, \ldots, V_k$ is isomorphic to any $V_{k+1}^\sigma, \ldots, V_n^\sigma$, for any $\sigma \in \mathfrak{G}$. Then the direct sum decomposition

$$V \underset{k}{\otimes} K = (V_1 \oplus \ldots \oplus V_k) \oplus (V_{k+1} \oplus \ldots \oplus V_n)$$

is stable under the action of $\mathfrak{G}$, and hence corresponds to a direct sum decomposition of V.

(vi) Since every KG-module is a direct summand of some $e_{k,K}(V)$ with V an indecomposable kG-module, it follows from (v) that $\mathrm{Ker}(f_{K,k})$ is the linear span of elements of the form $W - W^\sigma$ for W an indecomposable KG-module, and $\mathrm{Im}(e_{K,k})$ is the linear span of elements of the form $\sum_{\sigma \in \mathfrak{G}} W^\sigma$.

(vii) Suppose U and V are kG-modules with $U \underset{k}{\otimes} K = U_1 \oplus \ldots \oplus U_m$ and $V \underset{k}{\otimes} K = V_1 \oplus \ldots \oplus V_n$. By (vi), all the $\langle U_i, e_{k,K}\,\tau(V)\rangle$ are equal, and so

$$\begin{aligned}\langle U_i, e_{k,K}\,\tau(V)\rangle &= (1/m)\langle e_{k,K}(U), e_{k,K}\,\tau(V)\rangle \\ &= (1/m)\langle U, \tau(V)\rangle \quad \text{by (iv)}, \\ &= \begin{cases} d_U/m & \text{if } U \cong V \\ 0 & \text{otherwise.}\end{cases}\end{aligned}$$

On the other hand

$$\begin{aligned}\langle U_i, \tau e_{k,K}(V)\rangle &= \sum_{j=1}^{n} \langle U_i, \tau(V_j)\rangle \\ &= \begin{cases} d_{U_i}.(\text{no. of } V_j \text{ isomorphic to } U_i) & \text{if } U \cong V \\ 0 & \text{otherwise.}\end{cases}\end{aligned}$$

But $d_{U_i} = d_U/(m.(\text{no. of } V_j \text{ isomorphic to } U_i))$, since the extension K/k is separable. Thus $e_{k,K}\tau(V)$ and $\tau e_{k,K}(V)$ have the same inner product with every element of $A_K(G)$, and so by 2.18.5 they are equal. □

2.33.2 Proposition

Suppose V is a non-projective indecomposable kG-module with $V \otimes_k K = V_1 \oplus .. \oplus V_n$. Suppose $0 \to \Omega^2 V \to X_V \to V \to 0$ is the almost split sequence terminating in V. Then tensoring with K, we obtain the direct sum of the almost split sequences terminating in the V_i:

$$0 \to \Omega^2 V_1 \oplus .. \oplus \Omega^2 V_n \to X_{V_1} \oplus .. \oplus X_{V_n} \to V_1 \oplus .. \oplus V_n \to 0.$$

Proof

We have

$$\begin{aligned}\mathrm{Ext}^1_{kG}(V, \Omega^2 V) \otimes_k K &\cong \mathrm{Ext}^1_{KG}(V \otimes_k K, \Omega^2 V \otimes_k K) \\ &\cong \mathrm{Ext}^1_{KG}(V_1 \oplus .. \oplus V_n, \Omega^2 V_1 \oplus .. \oplus \Omega^2 V_n).\end{aligned}$$

The almost split sequence $0 \to \Omega^2 V \to X_V \to V \to 0$ corresponds to a generator x for the socle of $\mathrm{Ext}^1_{kG}(V, \Omega^2 V) \cong (\mathrm{End}_{kG}(V)/J(\mathrm{End}_{kG}(V)))^*$ Now $\mathrm{End}_{kG}(V)/J(\mathrm{End}_{kG}(V))$ is a division ring D, and without loss of generality we may take $x: D \to k$ to be the reduced trace function (i.e. tensor D with a splitting field, so that it becomes an algebra of matrices, and then x takes a matrix to its trace, which is in k). Then since K is a separable extension of k, $D \otimes_k K$ is a direct sum of complete matrix algebras over division rings with k in their centres (1.2.4) and $x \otimes_k K$ is still the reduced trace function. Thus as an element of $\mathrm{Ext}^1_{KG}(V_1 \oplus .. \oplus V_n, \Omega^2 V_1 \oplus .. \oplus \Omega^2 V_n)$, $x \otimes_k K$ represents the sum of the generators for the socles of $\mathrm{Ext}^1_{KG}(V_i, \Omega^2 V_i)$, and so our sequence is the direct sum of the almost split sequences. □

2.33.3 Proposition

Let q be a connected component of the stable quiver of kG-modules. Then the direct summands of $V \otimes_k K$ for $V \in q$ belong to a finite set of connected components $Q_1, \ldots, Q_n$ of the stable quiver of KG-modules, and σ acts transitively in the Q_i.

Proof

Choose a module $V \in q$, and let $V_1, \ldots, V_m$ be the isomorphism classes of summands of $V \otimes_k K$. By 2.33.2 and induction, if $W \in q$, and W_1 is a direct summand of $W \otimes_k K$, then W_1 is in the same connected component as one of the V_i. Thus there are at most m connected components., and by 2.33.2, $\mathfrak{G}$ acts transitively on them. □

Definitions

In the situation of 2.33.3, we say $Q_1, .., Q_n$ lie above q. If Q is a connected component of the stable quiver of KG-modules, we define the decomposition group $\mathfrak{G}_Q$ to be $\mathrm{Stab}_{\mathfrak{G}}(Q)$, and the decomposition field K^d to be the fixed field of $\mathfrak{G}_Q$.

2.33.4 Proposition

Let Q be a connected component of the stable quiver of KG-modules, with decomposition group $\mathfrak{G}_Q$ and decomposition field K^d. Let $Q_1, .., Q_n$ be the images of Q under $\mathfrak{G}$, and let q be the component of the stable quiver of kG-modules, over which they lie. Let $\mathfrak{q}$ be the component of the stable quiver of K^dG-modules over which Q lies. Then

(i) Q is the only component of the stable quiver of KG-modules lying over $\mathfrak{q}$,

(ii) there is a natural isomorphism $\mathfrak{q} \cong q$, and

(iii) $\mathfrak{q}$ is the quotient of Q by the action of $\mathfrak{G}_Q = \mathrm{Gal}(K/K^d)$.

Proof

(i) This is clear from the definition of K^d.

(ii) The isomorphism is given as follows. If $V \in q$, then $V \otimes_k K^d$ has a unique summand in $\mathfrak{q}$, since Gal(K/k) is transitive on the isomorphism classes of summands of $V \otimes_k K$ (2.33.1(v)), and $\mathfrak{G}_Q$ is precisely the setwise stabilizer of those summands lying in Q. The isomorphism in question takes V to this summand. This is clearly a quiver isomorphism by 2.33.2.

(iii) This is clear. □

Since $\mathfrak{G}_Q$ acts on Q, this passes down to an action of $\mathfrak{G}_Q$ on the reduced graph Δ of Q (but not on the associated tree, as we shall see).

The reduced graph Δ_o for $\mathfrak{q}$ may be obtained as follows. The vertices of Δ_o are the orbits of $\mathfrak{G}_Q$ on Δ. To find the new multiplicity a_{ij}, pick a representative i_o of the orbit i, and add together with a_{i_o,j_o} as j_o runs over the elements of the orbit j connected to i.

Example

Let $G = A_4$, $k = \mathbb{F}_2$ and $K = \mathbb{F}_4$. Let Q be the component of the stable quiver of KG-modules corresponding to the component of the Auslander-Reiten quiver containing the projective modules (see Appendix). Let $\mathfrak{q}$ be the corresponding component of the stable quiver of kG-modules. Then the tree class and reduced graph of Q are A_∞^∞ and $\tilde{A}_5$. The Galois group Gal(K/k) acts on $\tilde{A}_5$ as follows,

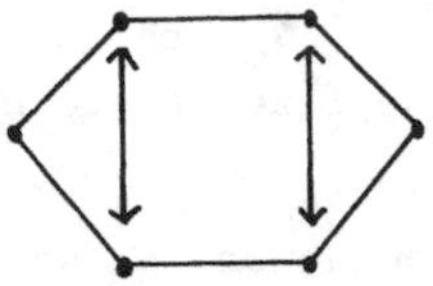

and so the reduced graph of $\mathcal{V}$ is

$\tilde{B}_3$ =  .

Since the tree class of $\mathcal{V}$ is also $\tilde{B}_3$, we see that the behaviour of the tree class under Galois descent is less easy to predict.

We define the <u>inertia group</u> T_Q of Q to be the pointwise stabilizer in $\mathfrak{G}_Q$ of the reduced graph of Q, and the <u>inertia field</u> K^t to be the fixed field of T_Q. The following proposition is clear.

2.33.5 <u>Proposition</u>

(i) The reduced graph of the component of the stable quiver of K^tG-modules corresponding to Q is isomorphic to the reduced graph of $\mathcal{V}$.

(ii) $\mathfrak{G}_Q/T_Q$ acts faithfully as a group of graph automorphisms on the reduced graph of Q, and is hence either cyclic or isomorphic to a subgroup of S_4. □

Appendix

Representations of particular groups

In this appendix, we list some information about the representation theory of particular finite groups. The amount of information given varies with the size of the group. We pay special attention to the representation theory of the Klein fours group, since this is a good example of many of the concepts introduced in the text. Our notation for the tables is a modification of the 'Atlas' conventions [36], as follows.

If A is a direct summand of A(G) satisfying hypothesis 2.21.1, we write first the atom table and then the representation table. The top row gives the value of c(s) (calculated using 2.21.13). The second row gives the ℓth power of s, for each relevant prime ℓ in numerical order (a prime is relevant for a species s if either $\ell \,\big|\, |\mathrm{Orig}(s)|$, or $p \,\big|\, |\mathrm{Orig}(s)|$ (p = char(k)) and $\ell \mid (p-1)$). The third row gives the isomorphism type of an origin of s, followed by a letter distinguishing the conjugacy class of the origin, and a number distinguishing the species with that origin, if there is more than one. Thus for example S3A2 means that the origin is isomorphic to S_3, and lies in a conjugacy class labelled 'A' ; the species in question is the second one with this origin. For the power maps (second row), the origin is determined by 2.16.11, so we only give the rest of the identifier.

The last column gives the conjugacy class of the vertex of the representation. If there is more than one possible source with a given vertex, the dimension of the source is given in brackets.

By 2.21.9, the Brauer character table of modular irreducibles always appears at the top left corner of the atom table. Similarly the Brauer character table of projective indecomposable modules always appears at the top left corner of the representation table.

For the irreducible modules in the atom table, we also give the Frobenius-Schur indicator, namely

+ if the representation is orthogonal
- if the representation is symplectic but not orthogonal
o if the representation is neither symplectic nor orthogonal.

(For char k ≠ 2, this is $(1, \psi^2(V))$, see example after 2.16.2).

Irrationalities

The irrationalities we find in these tables are as follows.

$$bn = \begin{cases} \frac{1}{2}(-1+\sqrt{n}) & \text{if } n \equiv 1 \pmod 4 \\ \frac{1}{2}(-1+i\sqrt{n}) & \text{if } n \equiv 3 \pmod 4 \end{cases}$$

i.e. the Gauss sum of half the primitive $n^{\underline{th}}$ roots of unity.

$zn = e^{2\pi i/n}$ is a primitive $n^{\underline{th}}$ root of unity

$yn = zn + \overline{zn}$

$rn = \sqrt{n}$

$in = i\sqrt{n}$

*m denotes the image of the adjacent irrationality under the Galois automorphism $zn \mapsto (zn)^m$

* denotes the conjugate of a quadratic irrationality.

** denotes *(-1).

x&m denotes x + x*m.

Projective modules

We give the Loewy structure of the projective indecomposable kG-modules. This is the diagram whose $i^{\underline{th}}$ row gives the simple summands of the ith Loewy layer, namely the completely reducible module $L_i(V) = VJ^{i-1}/VJ^i$, where $J = J(kG)$. In this diagram, simple modules are labelled by their dimensions, with some form of decoration (e.g. a subscript) if there is more than one simple module of the same dimension.

Auslander-Reiten Quiver

We use dotted lines to indicate the arrows involving projective modules, so that the stable quiver may be obtained by removing these arrows and the projective modules attached to them. When the tree class of a connected component is equal to the reduced graph, we only give the former. If they are different, we give both, and we write (tree class) → (reduced graph).

Cohomology

We give $H^*(G,\mathbb{Z})$ and $H^*(G,k)$ in the forms $\mathbb{Z}$[generators]/(relations) and k[generators]/(relations), where the relations $xy = (-1)^{\deg(x)\deg(y)}yx$ are to be assumed. We also give the Poincaré series $\xi_k(t) = \Sigma t^n \dim_k H^n(G,k)$.

Acknowledgement

I would like to thank Richard Parker for his permission to

reproduce extracts from his collection of decomposition matrices and Brauer character tables.

C_2 , the cyclic group of order two

i. Ordinary characters

	2	2
p power		A
ind	1A	2A
+	1	1
+	1	-1

$H^*(C_2,\mathbb{Z}) = \mathbb{Z}[x]/(2x)$, $\deg(x) = 2$.

ii. Representations over $\bar{\mathbb{F}}_2$ Representation type: finite.

Decomposition matrix and Cartan matrix

D =		I
	1	1
	1'	1

C =		I
	I	2

Atom table and representation table for A(G)

	2	-2
p power		A
ind	1A	2A
+	1	1
	0	-2

	2	-2	
p power		A	
	1A	2A	vtx
	2	0	1A
	1	1	2A

Projective indecomposable modules

I
I

Almost Split Sequences

$$0 \to I \to \begin{matrix} I \\ I \end{matrix} \to I \to 0$$

Auslander-Reiten Quiver

Tree Class A_1

Cohomology

$H^*(C_2,k) = k[x], \quad \deg(x) = 1, \quad xSq^1 = x^2 .$

$\xi_k(t) = \frac{1}{1-t}$

$\mathrm{Max}(H^{ev}(C_2,k)) \cong \mathbb{A}^1(k)$

$\mathrm{Proj}(H^{ev}(C_2,k))$ is a single point.

A Ring Homomorphism

If $H = \langle t \rangle$ is a cyclic subgroup of a group G with $|H| = 2$, then $(1+t)^2 = 0$, and

$$V \mapsto \mathrm{Ker}_V(1+t)/\mathrm{Im}_V(1+t)$$

is a ring homomorphism from $A(G)$ to $A(C_G(H)/H)$.

V_4, the Klein fours group $(= C_2 \times C_2)$

$H^*(V_4,\mathbb{Z}) = \mathbb{Z}[x,y,z]/(2x,2y,2z,z^2-xy^2-x^2y)$

$\deg(x) = \deg(y) = 2, \quad \deg(z) = 3.$

Representations over $\bar{\mathbb{F}}_2$ Representation type: tame.

The set of species of $A_k(V_4)$ falls naturally into three subsets.

(i) The dimension.

(ii) A set of species s_z parametrized by the non-zero complex numbers $z \in \mathbb{C}\backslash\{0\}$.

(iii) A set of species $s_{N,\lambda}$ parametrized by the set of ordered pairs (N,λ) with $N \in \mathbb{N}\backslash\{0\}$ and $\lambda \in \mathbb{P}^1(k)$.

The set of indecomposable representations also falls naturally into three subsets.

(i) The projective indecomposable representation P_1 of dimension four.

(ii) The syzygies of the trivial module $V_o = k$

$$V_m = \Omega^m(k) \quad \text{and} \quad V_{-m} = \mho^m(k).$$

(iii) The set of representations $V_{n,\lambda}$ parametrized by the set of ordered pairs (n,λ) with $n \in \mathbb{N}\backslash\{0\}$ and $\lambda \in \mathbb{P}^1(k)$, having dimension $2n$, and $\Omega(V_{n,\lambda}) \cong V_{n,\lambda}$.

Matrices for these representations are given as follows. Let $V_4 = \langle g_1,g_2 : g_1^2 = g_2^2 = 1,\ g_1g_2 = g_2g_1 \rangle$. Then $V_{n,\lambda}$ is the representation

$$g_1 \mapsto \begin{pmatrix} I & I \\ 0 & I \end{pmatrix} \qquad g_2 \mapsto \begin{pmatrix} I & J_\lambda \\ 0 & I \end{pmatrix}$$

where I represents an $n \times n$ identity matrix, while J_λ represents an $n \times n$ Jordan block with eigenvalue λ.

For $\lambda = \infty$, the representation is

$$g_1 \mapsto \begin{pmatrix} I & J_o \\ 0 & I \end{pmatrix} \qquad g_2 \mapsto \begin{pmatrix} I & I \\ 0 & I \end{pmatrix} .$$

$\mathrm{Aut}(V_4) = S_3 = < h_1, h_2 : h_1{}^2 = h_2{}^2 = (h_1h_2)^3 = 1,\ g_1^{h_1} = g_1,\ g_2^{h_1} = g_1g_2,\ g_1^{h_2} = g_2,\ g_2^{h_2} = g_1 >$ acts on the set of representations as follows. P_1 and V_n are fixed by al automorphisms, and

$$h_1 : V_{n,\lambda} \mapsto V_{n,1+\lambda} \quad , \quad h_2 : V_{n,\lambda} \mapsto V_{n,1/\lambda} \ .$$

Define infinite matrices A,B,C and D as follows

$\underset{\to}{N}$

A = $n\downarrow$

	1	2	3	4	5	..
1	2	0	0	0	0	
2	2	2	0	0	0	
3	2	2	2	0	0	
4	2	2	2	2	0	
5	2	2	2	2	2	
.						⋱
.						

$\underset{\to}{N}$

B = $n\downarrow$

	1	2	3	4	5	..
1	$\sqrt{2}$	$-\sqrt{2}$	0	0	0	
2	2	2	0	0	0	
3	2	2	2	0	0	
4	2	2	2	2	0	
5	2	2	2	2	2	
.						⋱
.						

$$C = n\downarrow \overset{N \rightarrow}{\begin{array}{c|cccccc} & 1 & 2 & 3 & 4 & 5 & \cdot\ \cdot \\ \hline 1 & -2 & 2 & 0 & 0 & 0 & \\ 2 & 0 & -2 & 2 & 0 & 0 & \\ 3 & 0 & 0 & -2 & 2 & 0 & \\ 4 & 0 & 0 & 0 & -2 & 2 & \\ 5 & 0 & 0 & 0 & 0 & -2 & \\ \cdot & & & & & & \ddots \\ \cdot & & & & & & \end{array}}$$

$$D = n\downarrow \overset{N \rightarrow}{\begin{array}{c|cccccc} & 1 & 2 & 3 & 4 & 5 & \cdot\ \cdot \\ \hline 1 & 2-2\sqrt{2} & 2+2\sqrt{2} & 0 & 0 & 0 & \\ 2 & \sqrt{2}-2 & -\sqrt{2}-2 & 2 & 0 & 0 & \\ 3 & 0 & 0 & -2 & 2 & 0 & \\ 4 & 0 & 0 & 0 & -2 & 2 & \\ 5 & 0 & 0 & 0 & 0 & -2 & \\ \cdot & & & & & & \ddots \\ \cdot & & & & & & \end{array}}$$

Let 0 represent an infinite matrix of zeros. Then the representation table and atom table for V_4 are as follows

<u>Representation Table for V_4 over $\overline{\mathbb{F}}_2$</u>

	Parameters	dim	z (s_z)	(N,∞) ($s_{N,\lambda}$)	$(N,0)$	$(N,1)$	(N,λ_1)	(N,λ_2)	(N,λ_3)	...
P_1	(projective)	4	0	0	0	0	0	0	0	...
V_m	m	$2\|m\|+1$	z^m	1	1	1	1	1	1	...
$V_{n,\lambda}$	(n,∞)	2n	0	A	0	0	0	0	0	...
	$(n,0)$	2n	0	0	A	0	0	0	0	...
	$(n,1)$	2n	0	0	0	A	0	0	0	...
	(n,λ_1)	2n	0	0	0	0	B	0	0	...
	(n,λ_2)	2n	0	0	0	0	0	B	0	...
	(n,λ_3)	2n	0	0	0	0	0	0	B	...
	⋮	⋮	⋮	⋮	⋮	⋮	⋮	⋮	⋮	⋱

Atom Table for V_4

Parameters	dim	z	(N,∞)	$(N,0)$	$(N,1)$	(N,λ_1)	(N,λ_1)	(N,λ_3)	...
(simple)	1	1	1	1	1	1	1	1	...
m	0	$-z^{m-1}(1-z)^2$	0	0	0	0	0	0	...
(n,∞)	0	0	C	0	0	0	0	0	...
$(n,0)$	0	0	0	C	0	0	0	0	...
$(n,1)$	0	0	0	0	C	0	0	0	...
(n,λ_1)	0	0	0	0	0	D	0	0	...
(n,λ_2)	0	0	0	0	0	0	D	0	...
(n,λ_3)	0	0	0	0	0	0	0	D	...
$\vdots$	$\vdots$	$\vdots$	$\vdots$	$\vdots$	$\vdots$	$\vdots$	$\vdots$	$\vdots$	$\ddots$

Direct Summands of A(G)

A(G,Cyc) is the linear span of P_1, $V_{1,0}$, $V_{1,1}$ and $V_{1,\infty}$.

A(G,Discrete) is the linear span of the elements P_1 and all the $V_{n,\lambda}$.

A_o(G,Discrete) is the linear span of elements of the form $2V_m - 2V_o - |m|P_1$, and is isomorphic to the ideal of $\mathbb{C}[X,X^{-1}]$ consisting of those functions vanishing at $X = 1$. Letting $\hat{A}_o$(G,Discrete) be the subring of A(G) generated by A_o(G,Discrete) and the identity element, we have

$$A(G) = A(G,\text{Discrete}) \oplus \hat{A}_o(G,\text{Discrete}) .$$

$\hat{A}_o$(G,Discrete) is isomorphic to $\mathbb{C}[X,X^{-1}]$, under the isomorphism $V_1 - \frac{1}{2}P_1 = X$, $V_{-1} - \frac{1}{2}P_1 = X^{-1}$. In particular, A(G,Discrete) $\oplus$ A_o(G,Discrete) is an ideal of codimension one in A(G), and A(G,Discrete) is not an ideal direct summand of A(G), since it has no identity element.

Power maps

$$(s_z)^m = \begin{cases} s_{z^m} & m \text{ odd} \\ s_z & m = 2 \end{cases}$$

$$(s_{N,\lambda})^m = \begin{cases} s_{N,\lambda} & m \text{ odd} \\ s_{N,\lambda^2} & m = 2 \end{cases}$$

<u>unless</u> $N = 1$ or 2, $\lambda \in \{0,1,\infty\}$ and $m \equiv 3$ or $5 \bmod 8$, in which case $(s_{1,\lambda})^m = s_{2,\lambda}$ and $(s_{2,\lambda})^m = s_{1,\lambda}$.

<u>Almost Split Sequences</u>

$$0 \to V_{n+2} \to V_{n+1} \oplus V_{n+1} \to V_n \to 0 \qquad (n \neq -1)$$

$$0 \to V_1 \to V_o \oplus V_o \oplus P_1 \to V_{-1} \to 0$$

$$0 \to V_{n,\lambda} \to V_{n+1,\lambda} \oplus V_{n-1,\lambda} \to V_{n,\lambda} \to 0 \qquad (n > 1)$$

$$0 \to V_{1,\lambda} \to V_{2,\lambda} \to V_{1,\lambda} \to 0$$

<u>Auslander-Reiten Quiver</u>

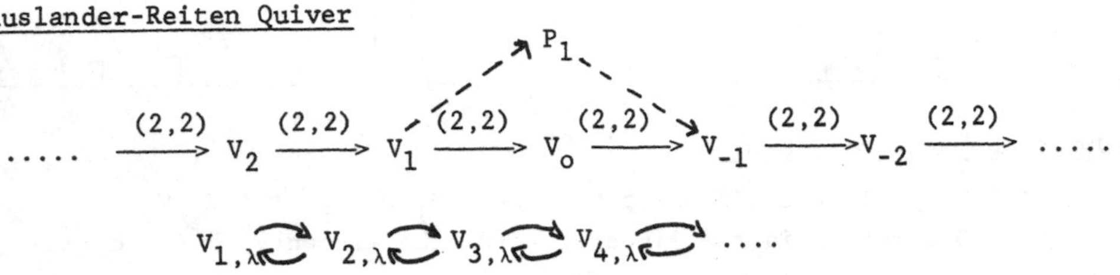

$(\lambda \in \mathbb{P}^1(k))$

<u>Tree Classes</u>

$\tilde{A}_{12}$, and a $\mathbb{P}^1(k)$- parametrized family of A_∞ 's.

<u>Cohomology</u>

$H^*(V_4,k) = k[x,y]$ $\qquad$ $\deg(x) = \deg(y) = 1$

$xSq^1 = x^2$, $\quad ySq^1 = y^2$

$\xi_k(t) = 1/(1-t)^2$

$\mathrm{Max}(H^{ev}(V_4,k)) \cong \mathbb{A}^2(k)$

$\mathrm{Proj}(H^{ev}(V_4,k)) \cong \mathbb{P}^1(k)$, and may be identified with the parametrizing set for the $V_{n,\lambda}$ in such a way that $\bar{\bar{X}}_G(V_{n,\lambda}) = \{\lambda\}$. Thus there is a one-one correspondence between the connected components of the stable quiver and the non-empty subvarieties of $\bar{X}_G$. Note that for a general group there may be many connected components of the stable quiver with the same variety.

Remark

For representation theory of V_4 over an arbitrary field in characteristic 2, see under the dihedral group of order 2^n.

The Dihedral Group D_{2^n}

$H^*(D_8,\mathbb{Z}) = \mathbb{Z}[w,x,y,z]/(2w,2x,2y,4z, y^2- wz, x^2- wx)$
$\deg(w) = \deg(x) = 2, \quad \deg(y) = 3, \deg(z) = 4.$

Representations over a field k of characteristic 2 (not necessarily algebraically closed) [79] Representation type: tame

First we describe the finite dimensional indecomposable modules for the infinite dihedral group

$$G = < x, y: x^2 = y^2 = 1> ,$$

and then we indicate which are modules for the quotient group $D_{4q} = < x, y: x^2 = y^2 = 1, \quad (xy)^q = (yx)^q > .$

Let $\mathbb{W}$ be the set of words in the letters a, b, a^{-1} and b^{-1} such that a and a^{-1} are always followed by b or b^{-1} and vice-versa, together with the 'zero length words' 1_a and 1_b. If C is a word, we define C^{-1} as follows. $(1_a)^{-1} = 1_b$, $(1_b)^{-1} = 1_a$; and otherwise, we reverse the order of the letters in the word and invert each letter according to the rule $(a^{-1})^{-1} = a$, $(b^{-1})^{-1} = b$. Let $\mathbb{W}_1$ be the set obtained from $\mathbb{W}$ by identifying each word with its inverse.

The nth power of a word of even length is obtained by juxtaposing n copies of the word. Let $\mathbb{W}'$ be the subset of $\mathbb{W}$ consisting of all words of even non-zero length which are not powers of smaller words. Let $\mathbb{W}_2$ be the set obtained from $\mathbb{W}'$ by identifying each word with its inverse and with its images under cyclic permutations

$$\ell_1 \cdot\cdot \ \ell_n \to \ell_n \ell_1 \cdot\cdot \ \ell_{n-1} \ .$$

The following is a list of all the isomorphism types of finite dimensional indecomposable kG-modules.

Modules of the first kind

These are in one-one correspondence with elements of $\mathbb{W}_1$. Let $C = \ell_1 \cdot\cdot \ \ell_n \in \mathbb{W}$. Let $M(C)$ be a vector space over k with basis $z_o , \cdot\cdot , z_n$ on which G acts according to the schema

$$kz_o \xleftarrow{\ \ell_1\ } kz_1 \xleftarrow{\ \ell_2\ } kz_2 \quad \cdot\cdot \quad kz_{n-1} \xleftarrow{\ \ell_n\ } kz_n$$

where x acts as " $1 + a$ " and y acts as " $1 + b$ ". (e.g. if $C = ab^{-1}aba^{-1}$ then the schema is

$$kz_o \xleftarrow{a} kz_1 \xrightarrow{b} kz_2 \xleftarrow{a} kz_3 \xleftarrow{b} kz_4 \xrightarrow{a} kz_5$$

and the representation is given by

$$x \mapsto \begin{pmatrix} 1 & & & & & \\ 1 & 1 & & & & \\ & & 1 & & & \\ & & 1 & 1 & & \\ & & & & 1 & 1 \\ & & & & & 1 \end{pmatrix} \quad y \mapsto \begin{pmatrix} 1 & & & & & \\ & 1 & 1 & & & \\ & & 1 & & & \\ & & & 1 & & \\ & & & 1 & 1 & \\ & & & & & 1 \end{pmatrix}.$$

It is clear that $M(C) \cong M(C^{-1})$.

Modules of the second kind

These are in one-one correspondence with elements of $\mathfrak{W}_2 \times \mathfrak{V}$ where

$\mathfrak{V} = \{(V,\varphi) :$ V is a vector space over k and φ is an indecomposable automorphism of V $\}$

(an indecomposable automorphism of a vector space is one whose rational canonical form has only one block, which is associated with a power of an irreducible polynomial over k). If $(C,(V,\varphi)) \in \mathfrak{W}' \times \mathfrak{V}$ with $C = \ell_1 \dots \ell_n$, let $M(C,V,\varphi)$ be the vector space $\bigoplus_{i=0}^{n-1} V_i$ with $V_i \cong V$ on which G acts according to the schema

$$V_o \xleftarrow{\ell_1 = \varphi} V_1 \xleftarrow{\ell_2 = id} V_2 \quad \dots \quad V_{n-2} \xleftarrow{\ell_{n-1} = id} V_{n-1}, \qquad V_o \xrightarrow{\ell_n = id} V_{n-1}$$

where again x acts as " $1 + a$ " and y acts as " $1 + b$ " as above. It is clear that if C and C' represent the same element of $\mathfrak{W}_2$ then $M(C,V,\varphi) \cong M(C',V,\varphi)$.

A module represents the quotient group D_{4q} if and only if either

(i) the module is of the first kind and the corresponding word does not contain $(ab)^q$, $(ba)^q$ or their inverses,

(ii) the module is of the second kind and no power of the corresponding word contains $(ab)^q$, $(ba)^q$ or their inverses, or

(iii) the module is the projective indecomposable module $M((ab)^q(ba)^{-q}, k, id)$ (of the second kind).

Almost Split Sequences

(a) Modules of the first kind

We define two functions L_q and R_q from words to words as follows. Let $A = (ab)^{q-1}a$ and $B = (ba)^{q-1}b$. If a word C starts with Ab^{-1} or Ba^{-1} then CL_q is obtained by cancelling that part; otherwise $CL_q = A^{-1}bC$ or $B^{-1}aC$ whichever is a word. Similarly if C ends in aB^{-1} or bA^{-1}, CR_q is obtained by cancelling that part; otherwise $CR_q = Ca^{-1}B$ or $Cb^{-1}A$, whichever is a word.

R_q and L_q are bijections from $\mathbb{W}$ to itself, and we have $R_qL_q = L_qR_q$, and $\Omega^2M(C) \cong M(CR_qL_q)$. The almost split sequence terminating in $M(C)$ is

$$0 \to M(CR_qL_q) \to M(CR_q) \oplus M(CL_q) \to M(C) \to 0$$

unless C or C^{-1} is AB^{-1}, in which case it is

$$0 \to M(CR_qL_q) \to M(CR_q) \oplus M(CL_q) \oplus P_1 \to M(C) \to 0$$

i.e.

$$0 \to M(A^{-1}B) \to M((ab)^{q-1}) \oplus M((ba)^{q-1}) \oplus M((ab)^q(ba)^{-q},k,id) \to M(AB^{-1}) \to 0$$

or $C = A$, in which case $0 \to M(A) \to M(Ab^{-1}A) \to M(A) \to 0$.

(b) Modules of the second kind

For an irreducible polynomial $p(x) \in k[x]$, let $(V_{n,p}, \varphi_{n,p})$ be the vector space and endomorphism with one rational canonical block associated with $(p(x))^n$. Then $\Omega^2M(C,V_{n,p},\varphi_{n,p}) \cong M(C,V_{n,p}, \varphi_{n,p})$ and the almost split sequence terminating in $M(C,V_{n,p}, \varphi_{n,p})$ is

$$0 \to M(C,V_{n,p},\varphi_{n,p}) \to M(C,V_{n+1,p},\varphi_{n+1,p}) \oplus M(C,V_{n-1,p}, \varphi_{n-1,p}) \to M(C,V_{n,p}, \varphi_{n,p}) \to 0 \qquad (n > 1)$$

and

$$0 \to M(C,V_{1,p}, \varphi_{1,p}) \to M(C,V_{2,p}, \varphi_{2,p}) \to M(C,V_{1,p}, \varphi_{1,p}) \to 0 .$$

Auslander-Reiten Quiver of D_{2^n}, $n \geq 3$

(a) Modules of the first kind

These fit together to form an infinite set of components of type $\mathbb{Z}A_\infty^\infty$

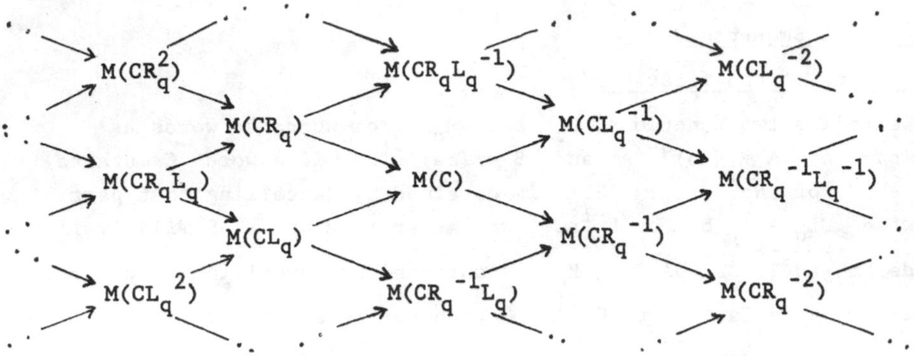

together with the following special components.

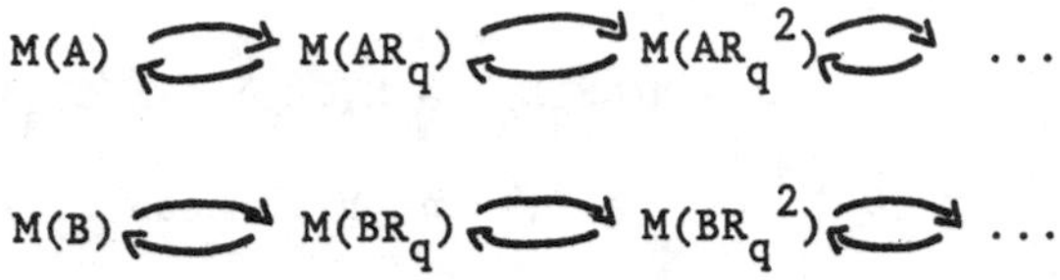

(note that $AR_qL_q = A$ and $BR_qL_q = B$)

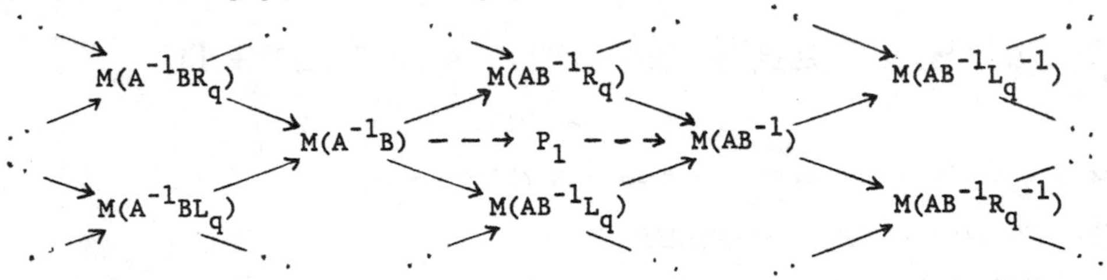

(note that $AB^{-1}R_qL_q = A^{-1}B$)

(b) <u>Modules of the second kind</u>

For each $C \in \mathbb{D}_2$ and each irreducible polynomial $p(x) \in k[x]$, there is a component

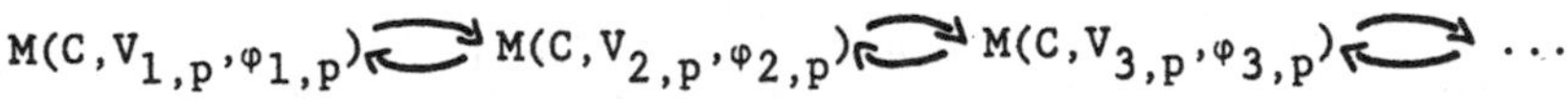

<u>Tree Classes</u>

(a) All components A_∞^∞.

(b) All components A_∞.

Cohomology of D_{2^n}, $n \geq 3$

$H^*(D_{2^n},k) = k[x,y,u]/(xy)$

	degree	Sq^1	Sq^2
x	1	x^2	0
y	1	y^2	0
u	2	$u(x+y)$	u^2

$\xi_k(t) = 1/(1-t)^2$

$\mathrm{Proj}(H^{ev}(D_{2^n},k))$ is a union of two copies of $\mathbb{P}^1(k)$ intersecting in a single point.

Varieties for Modules

Write $\overline{X}_G = \mathbb{P}^1_a \cup \mathbb{P}^1_b$, where $\mathbb{P}^1_a \cap \mathbb{P}^1_b = \{\infty_a\} = \{\infty_b\}$. Label in such a way that $\mathbb{P}^1_a$ corresponds to $<x,(xy)^q>$ and $\mathbb{P}^1_b$ corresponds to $<(xy)^q,y>$. By the Quillen stratification theorem (2.26.7) we have a homeomorphism

$$\overline{X}_{<x,(xy)^q>} \to \mathbb{P}^1_a .$$

Label $\overline{X}_{<x,(xy)^q>}$ in such a way that x corresponds to 0, $(xy)^q$ corresponds to ∞, and $(xy)^q x$ corresponds to 1, and write the homeomorphism as $\lambda \to \lambda(1+\lambda)$. Label $\mathbb{P}^1_b$ similarly with respect to $<(xy)^q,y>$. Then the varieties for the above modules are as follows.

$$\overline{X}_G(M(C)) = \begin{cases} \mathbb{P}^1_a \cup \mathbb{P}^1_b & \text{if } C \sim a^{\pm 1} \,..\, b^{\pm 1} \\ \mathbb{P}^1_b & \text{if } C \sim a^{\pm 1} \,..\, a^{\pm 1} \text{ but } C \not\sim (ab)^{q-1}a \\ \mathbb{P}^1_a & \text{if } C \sim b^{\pm 1} \,..\, b^{\pm 1} \text{ but } C \not\sim (ba)^{q-1}b \\ \{0_b\} & \text{if } C \sim (ab)^{q-1}a \\ \{0_a\} & \text{if } C \sim (ba)^{q-1}b \end{cases}$$

$$\overline{X}_G\left(M\left(C, \begin{pmatrix} \lambda 1 & & 0 \\ & \ddots & 1 \\ 0 & & \lambda \end{pmatrix}\right)\right) = \begin{cases} \{\infty\} & \text{unless } C \sim (ab)^{q-1}ab^{-1} \\ & \quad \text{or } C \sim (ba)^{q-1}ba^{-1} \\ \{\lambda_b\} & \text{if } C \sim (ab)^{q-1}ab^{-1} \\ \{\lambda_a\} & \text{if } C \sim (ba)^{q-1}ba^{-1} \end{cases} .$$

Q_8, the Quaternion group of order eight

$H^*(Q_8,\mathbb{Z}) = \mathbb{Z}[x,y,z]/(2x,2y,8z,x^2,y^2,xy - 4z)$

$\deg(x) = \deg(y) = 2$, $\deg(z) = 4$.

Representations over $\overline{\mathbb{F}}_2$

Representation type : tame

All modules are periodic with period 1, 2 or 4.

Auslander-Reiten Quiver

The tree class of each connected component of the stable quiver is A_∞.

Cohomology

$H^*(Q_8,k) = k[x,y,z]/(x^2 + xy + y^2, x^3, y^3)$

	degree	Sq^1	Sq^2	Sq^4
x	1	x^2	0	0
y	1	y^2	0	0
z	4	0	0	z^2

$\xi_k(t) = (1+t+t^2)/(1-t)(1+t^2)$

$\mathrm{Max}(H^{ev}(Q_8,k)) \cong \mathbb{A}^1(k)$

$\mathrm{Proj}(H^{ev}(Q_8,k))$ is a single point.

C_p, the cyclic group of order p, p odd

$H^*(C_p,\mathbb{Z}) = \mathbb{Z}[x]/(px)$, $\deg(x) = 2$.

Representations over $\mathbb{F}_p$ Representation type: finite.

Representation Table

There are p indecomposable representations X_j, $1 \le j \le p$, of dimension j, corresponding to the Jordan blocks with eigenvalue one. There are p species $s_o, \dots, s_{p-1}$ with

$$(s_k, X_j) = \frac{\sin(jk\pi/p)}{\sin(k\pi/p)}$$

$$c_k = c(s_k) = \frac{(-1)^k . p}{1 + \cos(k\pi/p)}$$

Power Maps $(s_k)^p = s_k$

$(s_k)^q = s_{qk}$ if $p \nmid q$

Almost split sequences

$$0 \to X_j \to X_{j+1} \oplus X_{j-1} \to X_j \to 0 \qquad 1 \le j \le p-1$$

Auslander -Reiten Quiver

$$X_1 \rightleftarrows X_2 \rightleftarrows \cdots \rightleftarrows X_{p-1} \rightleftarrows X_p$$

Tree class A_{p-1}

Cohomology

$H^*(C_p,k) = k[x,y]/(x^2)$ $\deg(x) = 1$, $\deg(y) = 2$, $x\beta = y$, $y\beta = 0$, $yP^1 = y^p$

$\xi_k(t) = \frac{1}{1-t}$

$\mathrm{Max}(H^{ev}(C_p,k)) \cong \mathbb{A}^1(k)$

$\mathrm{Proj}(H^{ev}(C_p,k))$ is a single point.

C_3 : atom table and representation table

	3	6	-2
p power		A1A1	A1A2
ind	1A	3A1	3A2
+	1	1	1
	0	3	-1
	0	-3	-1

	3	6	-2	
p power		A1A1	A1A2	
	1A	3A1	3A2	vtx
	3	0	0	1A
	1	1	1	3A(1)
	2	-1	1	3A(2)

The general p-group

Structure of the Group Algebra (Jennings, [61])

Let P be a p-group, and k a field of characteristic p. Define $H_1 = P$, and

$$H_i = \langle [H_{i-1}, P] , H^{(p)}_{(i/p)} \rangle$$

where (i/p) is the least integer which is greater than or equal to i/p, and $H_\lambda^{(p)}$ denotes the set of p-th powers of elements of H_λ. Then $\{H_i\}$ is minimal among series $\{G_i\}$ with

$$[G_i, P] \subseteq G_{i+1}$$

and $$x \in G_i \Rightarrow x^p \in G_{ip}.$$

In particular, H_i/H_{i+1} is elementary abelian, say of order p^{d_i}. Define $\phi(x) = 1 + x + .. + x^{p-1}$, and

$$F_P(x) = \phi(x)^{d_1} . \phi(x^2)^{d_2} . \quad .. \quad . \phi(x^m)^{d_m} = \Sigma a_i x^i$$

(m is the last value of i with $d_i \neq 0$). Then the dimension of the i+1-th Loewy layer of the group algebra (which is the only projective indecomposable module) is

$$\dim_k((J(kP))^i/(J(kP))^{i+1}) = a_i .$$

In particular, the Loewy length of kP is

$$\ell = 1 + \sum_i i.d_i(p-1) .$$

Since $F_P(x) = x^{\ell-1} F_P(1/x)$, we have $a_i = a_{\ell-1-i}$, and so the Loewy and Socle series of kP are the same.

<u>Groups of order p^2 and p^3</u> (p odd) [66]

$H^*(\mathbb{Z}/p^2\mathbb{Z},\mathbb{Z}) = \mathbb{Z}[x]/(p^2x)$, $\deg(x) = 2$.

$H^*(\mathbb{Z}/p\mathbb{Z} \times \mathbb{Z}/p\mathbb{Z},\mathbb{Z}) = \mathbb{Z}[x,y,z]/(px,py,pz,z^2)$

$\deg(x) = \deg(y) = 2$, $\deg(z) = 3$.

$H^*(\mathbb{Z}/p^3\mathbb{Z},\mathbb{Z}) = \mathbb{Z}[x]/(p^3x)$, $\deg(x) = 2$.

$H^*(\mathbb{Z}/p^2Z \times \mathbb{Z}/p\mathbb{Z},\mathbb{Z}) = \mathbb{Z}[x,y,z]/(p^2x,py,pz,z^2)$

$\deg(x) = \deg(y) = 2$, $\deg(z) = 3$.

$H^*(\mathbb{Z}/p\mathbb{Z} \times \mathbb{Z}/p\mathbb{Z} \times \mathbb{Z}/p\mathbb{Z},\mathbb{Z}) = \mathbb{Z}[x_1,x_2,x_3,y_1,y_2,y_3,z]/(px_1,px_2,px_3,py_1,py_2,py_3,pz,y_1^2,y_2^2,y_3^2,z^2,y_1z,y_2z,y_3z,y_2y_3+x_1z,y_1y_3+x_2z,y_1y_2+x_3z,x_1y_1+x_2y_2 + x_3y_3)$

$\deg(x_1) = \deg(x_2) = \deg(x_3) = 2$, $\deg(y_1) = \deg(y_2) = \deg(y_3) = 3$, $\deg(z) = 4$.

If $G = \langle g,h : g^{p^2} = h^p = 1, h^{-1}gh = g^{1+p} \rangle$ then

$H^*(G,\mathbb{Z}) = \mathbb{Z}[w,x,y,z_1,\ldots,z_{p-1}]/(pw,px,p^2y,pz_i,x^2,wz_i,xz_i,z_iz_j)$

$\deg(w) = 2$, $\deg(x) = 2p+1$, $\deg(y) = 2p$, $\deg(z_i) = 2i$.

If $G = \langle g,h,k\colon g^p = h^p = k^p = 1,\ [g,h] = k,\ [g,k] = [h,k] = 1\rangle$ then

$H^*(G,\mathbb{Z}) = \mathbb{Z}[x_1,x_2,x_3,x_4,x_5,x_6,y_1,\ldots,y_{p-3}]/(px_1,px_2,px_3,px_4,px_5,p^2x_6,$
$py_i,x_3^2,x_4^2,x_1y_i,x_2y_i,x_3y_i,x_4y_i,x_5y_i,y_iy_j,x_1x_3-x_2x_4,x_1^px_3-x_2^px_4,$
$x_1^px_2-x_2^px_1,x_5^p-x_1^{p-1}x_2^{p-1},x_1(x_5-x_2^{p-1}),x_2(x_5-x_1^{p-1}),x_3(x_5-x_1^{p-1}),$
$x_4(x_5-x_2^{p-1}))$

$\deg(x_1) = \deg(x_2) = 2$, $\deg(x_3) = \deg(x_4) = 3$, $\deg(x_5) = 2p-2$,
$\deg(x_6) = 2p$, $\deg(y_i) = 2i+2$.

The p-hypoelementary groups with cyclic O_p $\qquad$ $k = \mathbb{F}_p$

Representation type: finite.

Let $H = \langle x,y\colon x^{p^r} = y^m = 1,\ x^y = x^a\rangle$ where a is a primitive d^{th} root of unity modulo p^r, d divides $p-1$ and d divides m. Let θ be a primitive m^{th} root of unity in k with $a \equiv \theta^{m/d}$ as elements of the prime field $\mathbb{F}_p$. There are m irreducible modules $X_1(\theta^q)$, $1 \le q \le m$, for H, which are one dimensional and are given by $x \mapsto (1)$, $y \mapsto (\theta^q)$ as matrices. If $1 \le n \le p^r$, there are m indecomposable modules of dimension n. These are denoted $X_n(\theta^q)$, $1 \le q \le m$. These account for all the irreducible modules. $X_n(\theta^q)$ is uniserial, with Loewy layers $L_i(X_n(\theta^q)) \cong X_1(a^{n-i}\,\theta^q)$. We write X_n for $X_n(1)$.

The case $r = 1$

In this case H has order $p.m$, $(p,m) = 1$. The following relations are sufficient to determine the structure of $A(G)$.

$$X_1(\theta^q) \otimes X_n \cong X_n(\theta^q)$$

$$X_2 \otimes X_n \cong \begin{cases} X_{n-1}(a) \oplus X_{n+1} & \text{if } 1 \le n < p \\ X_p(a) \oplus X_p & \text{if } n = p\,. \end{cases}$$

Let λ be a primitive $2m^{th}$ root of unity in $\mathbb{C}$. Then the representation table is as follows.

	Brauer species		Non-Brauer species
Species / Rep^n	dim	${}^b y^t \quad (1 \le t < m)$	$S_{\pm t_1, t} \quad (1 \le t_1 < p,\ 0 \le t < m)$
$X_n(\theta^q)$	n	$\lambda^{2qt}\left(\frac{1-\lambda^{2mnt/d}}{1-\lambda^{2mt/d}}\right)$	$\lambda^{2qt+t(n-1)m/d}\ \frac{\sin(nt_1\pi/p)}{\sin(t_1\pi/p)}$

Almost split sequences (general r)

$$0 \to X_j(\theta^q) \to X_{j+1}(\theta^q) \oplus X_{j-1}(a\,\theta^q) \to X_j(a\,\theta^q) \to 0$$

$$1 \le j < p^r, \quad 0 \le q < m$$

Auslander-Reiten Quiver

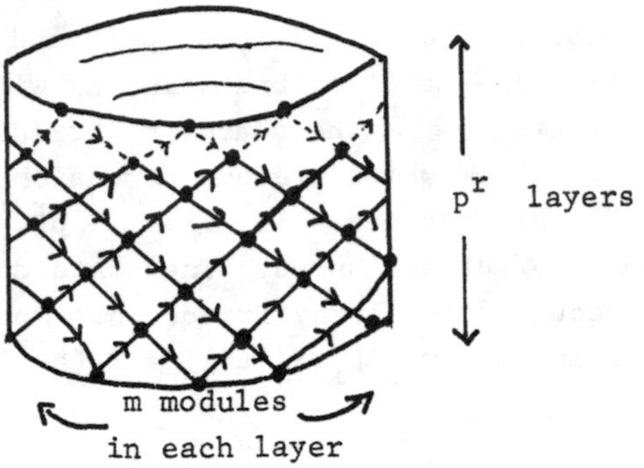

The stable quiver is obtained by deleting the top layer, which consists of the modules $X_{p^r}(\theta^q)$, and all the arrows connected to them.

Tree class $\quad A_{p^r-1}$

Cohomology $\quad (p^r \neq 2)$

$H^*(H,k) = k[x,y]/(x^2) \qquad \deg(x) = 1, \quad \deg(y) = 2$

$x\beta = y, \ y\beta = 0, \ yP^1 = y^p.$

$\xi_k(t) = 1/(1-t)$

$\mathrm{Max}(H^{ev}(H,k)) \cong \mathbb{A}^1(k)$

$\mathrm{Proj}(H^{ev}(H,k))$ is a single point.

S_3: atom table and representation table, $k = \bar{\mathbb{F}}_3$

	6	2	12	-4	-4	-4
p power		A	A1A1	A1A2	A1A1	A1A2
ind	1A	2A	3A1	3A2	S3A1	S3A2
+	1	1	1	1	1	1
+	1	-1	1	1	-1	-1
	0	0	3	-1	-1	-1
	0	0	3	-1	1	1
	0	0	-3	-1	i	-i
	0	0	-3	-1	-i	i

	6	2	12	-4	-4	-4	
p power		A	A1A1	A1A2	A1A1	A1A2	
	1A	2A	3A1	3A2	S3A1	S3A2	vtx
	3	1	0	0	0	0	1A
	3	-1	0	0	0	0	1A
	1	1	1	1	1	1	3A(1)
	1	-1	1	1	-1	-1	3A(1)
	2	0	-1	1	-i	i	3A(2)
	2	0	-1	1	i	-i	3A(2)

The Alternating Group A_4

i. Ordinary Characters

We display in one table, according to 'Atlas' conventions, the ordinary characters of A_4, $2A_4$, S_4 and $2S_4$. Note that there are two isomorphism classes of isoclinic groups $2S_4$, and to get from the character table of one to the character table of the other we multiply the bottom right hand corner by i.

	12	4	3	3			2	2
p power		A	A	A			A	A
ind	1A	2A	3A	B**	fus	ind	2B	4A
+	1	1	1	1	:	++	1	1
o	1	1	z3	**		+	0	0
o	1	1	**	z3				
+	3	-1	0	0	:	++	1	-1
ind	1 2	4	3 6	3 6	fus	ind	2	8 8
-	2	0	-1	-1	:	oo	0	i2
o	2	0	-z3	**		+	0	0
o	2	0	**	-z3				

ii. Representations over $\overline{\mathbb{F}}_2$

Representation type: tame.

Write ω, $\bar{\omega}$ for the primitive cube roots of unity in both k and $\mathbb{C}$.

Let $A_4 = \langle g_1, g_2, h : g_1{}^2 = g_2{}^2 = h^3 = 1,\ g_1g_2 = g_2g_1,\ g_1{}^h = g_2,\ g_2{}^h = g_1g_2\rangle$. Let h act on $\mathbb{P}^1(k)$, the parametrizing set for representations and species for V_4, via $h : \lambda \mapsto 1/(1+\lambda)$. Then $V_{n,\lambda}\!\uparrow^{A_4} \cong V_{n,\mu}\!\uparrow^{A_4}$ if and only if λ and μ represent the same element of $\mathbb{P}_1(k)/\langle h\rangle$.

The indecomposable representations of A_4 are obtained by taking direct summands of representations induced up from V_4, and are as follows.

(i) The projective covers P_1, P_ω, $P_{\bar{\omega}}$ of the simple modules 1, ω and $\bar{\omega}$.

(ii) $W_n(\alpha) = \Omega^n(\alpha)$, $\alpha \in \{1, \omega, \bar{\omega}\}$, $n \in \mathbb{Z}$.

(iii) $W_{n,\lambda} = V_{n,\lambda}\!\uparrow^{A_4}$, $\lambda \in (\mathbb{P}^1(k)/\langle h\rangle)\setminus\{\omega, \bar{\omega}\}$, $n \in \mathbb{N}\setminus\{0\}$.

(iv) $W_{n,\omega}(\alpha)$, $W_{n,\bar{\omega}}(\alpha)$, $\alpha \in \{1, \omega, \bar{\omega}\}$, $n \in \mathbb{N}\setminus\{0\}$.

These last representations, of dimension $2n$, ,are the direct summands of $V_{n,\omega}\!\uparrow^{A_4}$ and $V_{n,\bar{\omega}}\!\uparrow^{A_4}$. They are defined as follows.

$$\left.\begin{aligned} W_{n,\omega}(\alpha) &= W_{n,\omega}(1) \otimes \alpha \\ W_{n,\bar{\omega}}(\alpha) &= W_{n,\bar{\omega}}(1) \otimes \alpha \end{aligned}\right\}\ \alpha \in \{1, \omega, \bar{\omega}\}$$

$$W_{n,\bar{\omega}}(1) = \psi^2(W_{n,\omega}(1)),$$

and $W_{n,\omega}(1)$ is given in terms of matrices as follows.

$$g_1 \mapsto \begin{pmatrix} I & \begin{matrix} \bar{\omega} & & & 0 \\ \omega & \bar{\omega} & & \\ & \omega & \ddots & \\ & & \ddots & \bar{\omega} \\ 0 & & \omega & \bar{\omega} \end{matrix} \\ O & I \end{pmatrix} \qquad g_2 \mapsto \begin{pmatrix} I & \begin{matrix} \omega & & & 0 \\ \bar{\omega} & \omega & & \\ & \bar{\omega} & \ddots & \\ & & \ddots & \omega \\ 0 & & \bar{\omega} & \omega \end{matrix} \\ O & I \end{pmatrix}$$

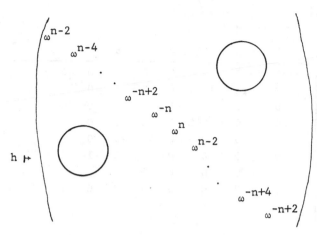

$$h \mapsto \begin{pmatrix} \omega^{n-2} & & & & & & & & & \\ & \omega^{n-4} & & & & & & \bigcirc & & \\ & & \cdot & & & & & & & \\ & & & \cdot & & & & & & \\ & & & & \omega^{-n+2} & & & & & \\ & & & & & \omega^{-n} & & & & \\ & & & & & & \omega^{n} & & & \\ & \bigcirc & & & & & & \omega^{n-2} & & \\ & & & & & & & & \cdot & \\ & & & & & & & & & \cdot \\ & & & & & & & & & \omega^{-n+4} \\ & & & & & & & & & & \omega^{-n+2} \end{pmatrix}$$

Tensor products modulo projectives are as follows. If $m \le n$,

$$W_{m,\omega}(1) \otimes W_{n,\omega}(1) \cong \begin{cases} W_{2,\omega}(1) & \text{if } m = n = 1 \\ 2.W_{m,\omega}(1) & \text{if } n \equiv 2 \pmod 3 \\ W_{m,\omega}(\omega) \oplus W_{m,\omega}(\bar{\omega}) & \text{otherwise.} \end{cases}$$

$$W_{m,\omega}(1) \otimes W_{n,\bar{\omega}}(1) = 0.$$

The species of $A_k(A_4)$ are as follows. There is one species for each h-orbit on species of V_4, there are two Brauer species corresponding to $\langle h \rangle$, and there are the species whose origin is A_4, namely

(i) two sets of species s_z and s_z^2 parametrized by the complex numbers $z \in \mathbb{C}\setminus\{0\}$, and

(ii) four sets of species $s_{n,\omega}$, $s_{n,\omega}^2$, $s_{n,\bar{\omega}}$, and $s_{n,\bar{\omega}}^2$.

The representation table and atom table are as follows.

Let A, B, C and D be as in the representation table and atom table of V_4, and define further infinite matrices E and F as follows.

E = (rows $n\downarrow$, columns $N\rightarrow$)

n \ N	1	2	3	4	5	6	. .
1	r2	-r2	0	0	0	0	
2	2	2	0	0	0	0	
3	-1	-1	-1	0	0	0	
4	-1	-1	-1	-1	0	0	
5	2	2	2	2	2	0	
6	-1	-1	-1	-1	-1	-1	
.							⋱
.							

(each column repeats with period three where it is non-zero)

$F = \; n \downarrow$ (rows n), columns $\underset{\rightarrow}{N}$:

	1	2	3	4	5	6	7	8	. .
1	2+r2	2-r2	0	0	0	0	0	0	
2	1+r2	1-r2	-1	0	0	0	0	0	
3	0	0	-2	-1	0	0	0	0	
4	0	0	0	1	2	0	0	0	
5	0	0	0	0	1	-1	0	0	
6	0	0	0	0	0	-2	-1	0	
7	0	0	0	0	0	0	1	2	
8	0	0	0	0	0	0	0	1	
.									
.									

We use the 'Atlas' format to make clear the relationship with the tables for V_4.

Representation Table

Parameters	dim	z	(N,∞)	$(N,0)$	$(N,1)$	(N,ω)	$(N,\bar\omega)$	(N,λ)	(N,λ^h)	(N,λ^{h^2})	fus	h	z	(N,ω)	$(N,\bar\omega)$
(projective)	4	0	0	0	0	0	0	0	0	0	:	1	0	0	0
m	$2\|m\|+1$	z^m	1	1	1	1	1	1	1	1	:	ε	z^m	1	1
(n,∞)	2n	0	A	0	0							0	0	0	0
$(n,0)$	2n	0	0	A	0	0			0						
$(n,1)$	2n	0	0	0	A										
(n,ω)	2n	0		0		B	0		0		:	ε	0	E	0
$(n,\bar\omega)$	2n	0				0	B				:	ε	0	0	E
(n,μ)	2n	0						$B\delta_{\lambda\mu}$	0	0		0	0	0	0
(n,μ^h)	2n	0		0		0		0	$B\delta_{\lambda\mu}$	0					
(n,μ^{h^2})	2n	0						0	0	$B\delta_{\lambda\mu}$					

Atom Table

Parameters	dim	z	(N,∞)	$(N,0)$	$(N,1)$	(N,ω)	$(N,\bar\omega)$	(N,λ)	(N,λ^h)	(N,λ^{h^2})	fus	h	z	(N,ω)	$(N,\bar\omega)$
(simple)	1	1	1	1	1	1	1	1	1	1	:	1	1	1	1
m	0	$-z^{m-1}(1-z)^2$	0	0	0	0	0	0	0	0	:	0	$-z^{m-1}(1-z)^2$	0	0
(n,∞)	0	0	C	0	0							0	0	0	0
$(n,0)$	0	0	0	C	0	0			0						
$(n,1)$	0	0	0	0	C										
(n,ω)	0	0		0		D	0		0		:	0	0	F	0
$(n,\bar\omega)$	0	0				0	D				:	0	0	0	F
(n,μ)	0	0						$D\delta_{\lambda\mu}$	0	0		0	0	0	0
(n,μ^h)	0	0		0		0		0	$D\delta_{\lambda\mu}$	0					
(n,μ^{h^2})	0	0						0	0	$D\delta_{\lambda\mu}$					

$m \in \mathbb{Z}$, $n \geq 1$,
$\varepsilon \in \{0, \pm 1\}$, $\varepsilon \equiv$ dimension (mod 3)
λ and μ run over a set of representatives in $\mathbb{P}^1(k)\setminus\{0,1,\infty,\omega,\bar{\omega}\}$ under the action of h, and $\delta_{\lambda\mu}$ represents the matrix which is the identity if $\lambda = \mu$ and zero otherwise.

Almost Split Sequences

$$0 \to W_{n+2}(\alpha) \to W_{n+1}(\alpha\omega) \oplus W_{n+1}(\alpha\bar{\omega}) \to W_n(\alpha) \to 0$$

$$0 \to W_1(\alpha) \to W_o(\alpha\omega) \oplus W_o(\alpha\bar{\omega}) \oplus P_\alpha \to W_{-1}(\alpha) \to 0$$

$$\alpha \in \{1, \omega, \bar{\omega}\}, \quad n \in \mathbb{Z}\setminus\{-1\}.$$

$$0 \to W_{n,\omega}(\alpha\omega) \to W_{n+1,\omega}(\alpha\bar{\omega}) \oplus W_{n-1,\omega}(\alpha\bar{\omega}) \to W_{n,\omega}(\alpha) \to 0$$

$$0 \to W_{1,\omega}(\alpha\omega) \to W_{2,\omega}(\alpha\bar{\omega}) \to W_{1,\omega}(\alpha) \to 0$$

$$0 \to W_{n,\bar{\omega}}(\alpha\bar{\omega}) \to W_{n+1,\bar{\omega}}(\alpha\omega) \oplus W_{n-1,\bar{\omega}}(\alpha\omega) \to W_{n,\bar{\omega}}(\alpha) \to 0$$

$$0 \to W_{1,\bar{\omega}}(\alpha\bar{\omega}) \to W_{2,\bar{\omega}}(\alpha\omega) \to W_{1,\bar{\omega}}(\alpha) \to 0$$

$$\alpha \in \{1, \omega, \bar{\omega}\}, \quad n > 1.$$

$$0 \to W_{n,\lambda} \to W_{n+1,\lambda} \oplus W_{n-1,\lambda} \to W_{n,\lambda} \to 0$$

$$0 \to W_{1,\lambda} \to W_{2,\lambda} \to W_{1,\lambda} \to 0$$

$$\lambda \in (\mathbb{P}^1(k)/\langle h\rangle) \setminus \{\omega,\bar{\omega}\}, \quad n > 1.$$

Auslander-Reiten Quiver

(a)

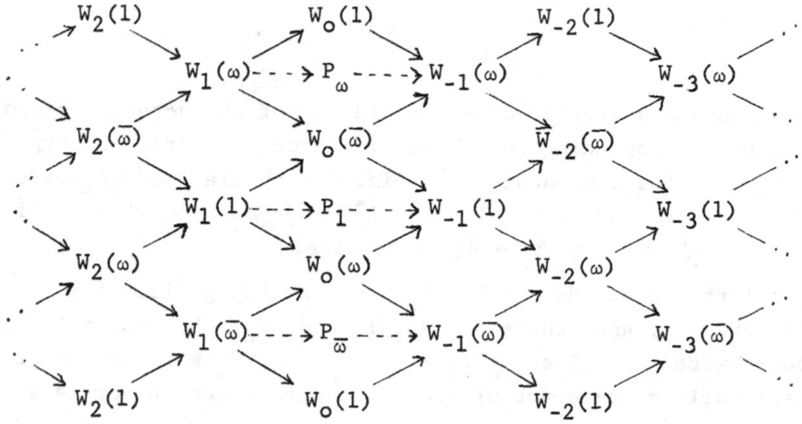

(Identify top and bottom lines to form a doubly infinite cylinder)

(b) Two connected components, for $\alpha = \omega$ and $\alpha = \bar{\omega}$ as follows.

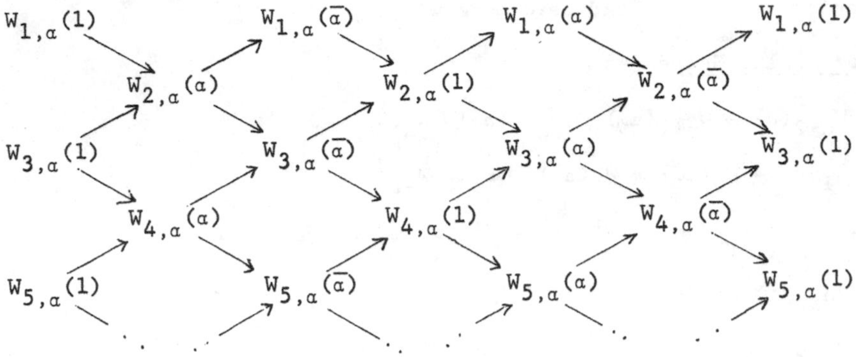

(Identify right hand and left hand edges to form a singly infinite cylinder)

(c) For each $\lambda \in (\mathbb{P}^1(k)/\langle h\rangle)\setminus\{\omega, \bar{\omega}\}$, a connected component as follows.

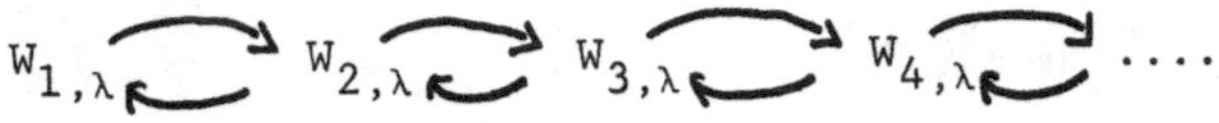

Tree Classes and Reduced Graphs

(a) $A_\infty^\infty \longrightarrow \tilde{A}_5$

(b) Two copies of $A_\infty \longrightarrow A_\infty$

(c) A $(\mathbb{P}^1(k)/\langle h\rangle)\setminus\{\omega, \bar{\omega}\}$ - parametrized family of copies of $A_\infty \longrightarrow A_\infty$.

Remark

Over a non algebraically closed field k of characteristic two, the components corresponding to (b) and (c) above are still of type $A_\infty \longrightarrow A_\infty$, but the component containing the trivial module (i.e. corresponding to (a) above) is of type $A_\infty^\infty \longrightarrow \tilde{A}_5$ if $x^2 + x + 1$ splits in k and of type $\tilde{B}_3 \to \tilde{B}_3$ otherwise.

In the latter case, denote by k_1 the splitting field for $x^2 + x + 1$ over k, and denote by $W_n(2)$ and P_2 the modules which when tensored with k_1 give $W_n(\omega) \oplus W_n(\bar{\omega})$ and $P_\omega \oplus P_{\bar{\omega}}$ respectively. Then the appropriate component of the Auslander-Reiten quiver is as follows.

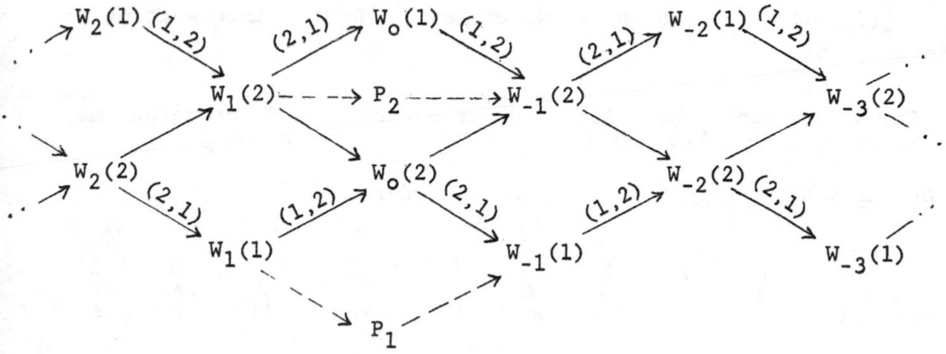

(Do not make any identifications)

<u>Cohomology of A_4 over $\mathbb{F}_2$</u>

$H^*(A_4,k)$ is the set of fixed points of h on $H^*(V_4,k) = k[x,y]$, namely the subring generated by $u = x^2 + xy + y^2$, $v = x^3 + x^2y + y^3$ and $w = x^3 + xy^2 + y^3$. Thus

$H^*(A_4,k) = k[u,v,w]/(u^3 + v^2 + vw + w^2)$

	degree	Sq^1	Sq^2
u	2	v+w	u^2
v	3	u^2	uw
w	3	u^2	uv

$\xi_k(t) = (1-t+t^2)/(1-t)^2(1+t+t^2)$

$\mathrm{Proj}(H^{ev}(A_4,k))$ is the irreducible conic in $\mathbb{P}^2(k)$ given by $x_1^2 = x_2x_3$.

<u>Cohomology of S_4 over $\mathbb{F}_2$</u>

$H^*(S_4,k) = k[x,y,z]/(xz)$

	degree	Sq^1	Sq^2	restriction to A_4
x	1	x^2	0	0
y	2	z+xy	y^2	u
z	3	xz	yz	v+w

$\xi_k(t) = (1+t^2)/(1-t)^2(1+t+t^2)$

$Proj(H^{ev}(S_4,k))$ is a union of two copies of $\mathbb{P}^1(k)$ intersecting in a single point.

The Alternating Group A_5, and its coverings and automorphisms.

i. Ordinary Characters (see remarks under A_4)

	60	4	3	5	5			6	2	3
p power		A	A	A	A			A	A	AB
ind	1A	2A	3A	5A	B*	fus	ind	2B	4A	6A
+	1	1	1	1	1	:	++	1	1	1
+	3	-1	0	-b5	*	┃	+	0	0	0
+	3	-1	0	*	-b5	┃				
+	4	0	1	-1	-1	:	++	2	0	-1
+	5	1	-1	0	0	:	++	1	-1	1
ind	1	4	3	5	5	fus	ind	2	8	6
	2		6	10	10				8	6
-	2	0	-1	b5	*	┃	-	0	0	0
-	2	0	-1	*	b5	┃				
-	4	0	1	-1	-1	:	oo	0	0	i3
-	6	0	0	1	1	:	oo	0	i2	0

ii. Representations over $\overline{\mathbb{F}}_2$ Representation type: tame

Decomposition matrix

	I	2_1	2_2	4
1	1	0	0	
3_1	1	1	0	
3_2	1	0	1	
4				1
5	1	1	1	

Cartan Matrix

	I	2_1	2_2	4
I	4	2	2	
2_1	2	2	1	
2_2	2	1	2	
4				1

Atom Table and Representation Table for A(G,Cyc)

	60 p power ind 1A	3 A 3A	5 A 5A1	5 A 5A2	-4 A 2A
+	1	1	1	1	1
-	2_1	-1	b5	*	0
-	2_2	-1	*	b5	0
+	4	1	-1	-1	0
	0	0	0	0	-2

60 p power 1A	3 A 3A	5 A 5A1	5 A 5A2	-4 A 2A	vtx
12	0	2	2	0	1A
8	-1	*	b5	0	1A
8	-1	b5	*	0	1A
4	1	-1	-1	0	1A
6	0	1	1	2	2A

Representation table for A(G,Triv), G = A_5, k = $\overline{\mathbb{F}}_2$

p power 1A	A 3A	A 5A	A B*	A 2A	A V4A	BA A4A	AA B**	vtx
12	0	2	2	0	0	0	0	1A
8	-1	b5	*	0	0	0	0	1A
8	-1	*	b5	0	0	0	0	1A
4	1	-1	-1	0	0	0	0	1A
6	0	1	1	2	0	0	0	2A
5	-1	0	0	1	1	z3	**	V4A
5	-1	0	0	1	1	**	z3	V4A
1	1	1	1	1	1	1	1	V4A

Projective Indecomposable Modules for A_5 over $\overline{\mathbb{F}}_2$

I		2_1	2_2	
2_1	2_2	I	I	
I	I	2_2	2_1	4
2_2	2_1	I	I	
I		2_1	2_2	

Projective Indecomposable modules for S_5 over $\mathbb{F}_2$

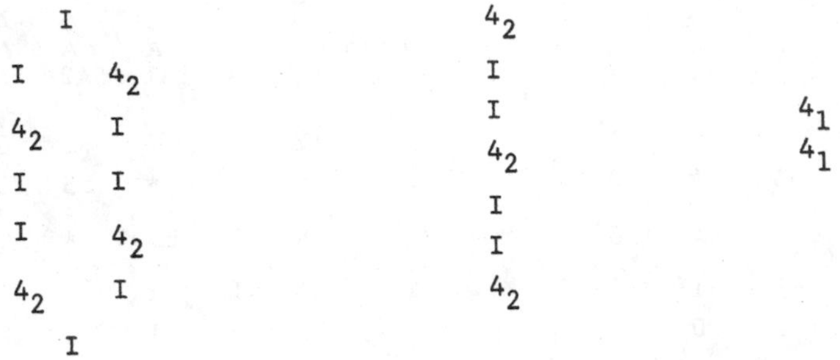

Projective Indecomposable modules for $2A_5$ over $\bar{\mathbb{F}}_2$

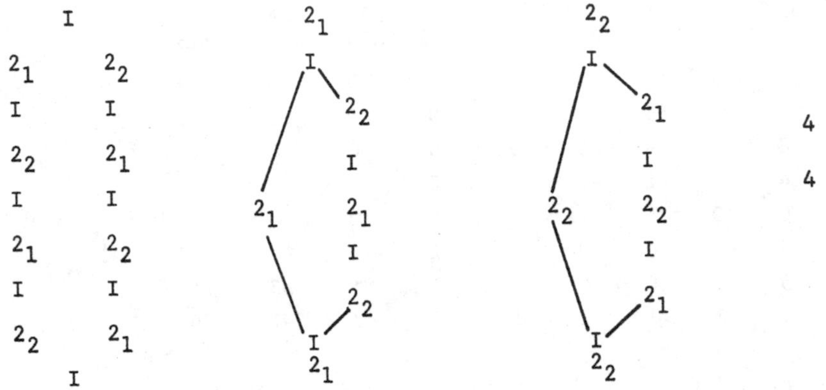

Green Correspondence between A_4 and A_5

Since a Sylow 2-subgroup of A_5 is a t.i. subgroup with normalizer A_4 (see exercise to 2.12), Green correspondence sets up a one-one correspondence between non-projective modules for A_5 and for A_4 in characteristic two. This correspondence takes almost split sequences to almost split sequences, and so the stable quivers are isomorphic. The atom copying theorem (section 2.20) gives the atom table as follows.

Atom Table

Parameters	1A	3A	5A1	5A2	z_j	(N,∞)	$(N,\omega)_j$	$(N,\bar{\omega})_j$	(N,λ)
trivial	1	1	1	1	1	1	1	1	1
$L_2(4)$-natural	2_1	-1	b5	*	0	0	G	0	0
dual	2_2	-1	*	b5	0	0	0	G	0
Steinberg	4	1	-1	-1	0	0	0	0	0
m_i					$-z^{m-1}(1-z)^2\omega^{i-j}$	0	0	0	0
(n,∞)					0	C	0	0	0
$(n,\omega)_i$	◯				0	0	$\begin{cases} D & i=j \\ F & i\neq j \end{cases}$	0	0
$(n,\bar{\omega})_i$					0	0	0	$\begin{cases} D & i=j \\ F & i\neq j \end{cases}$	0
(n,μ)					0	0	0	0	$D\delta_{\lambda\mu}$

In this table, D and F are as given under V_4 and A_4, and

$$G = (r2,\ -r2, 0, 0,\ \ldots)$$

$$i \in \{1, \omega, \bar{\omega}\}$$

λ and μ run over a set of representatives in $\mathbb{P}^1(k)\setminus\{0,1,\infty,\omega,\bar{\omega}\}$ under the action of $t \mapsto 1/(1-t)$, and $\delta_{\lambda\mu}$ represents the matrix which is the identity if $\lambda = \mu$ and zero otherwise.

Almost split sequences for A_5 over $\bar{\mathbb{F}}_2$

These are given by applying Green correspondence to the almost split sequences for A_4. The sequences involving projective modules are as follows.

$$0 \to g(W_1(1)) \to P_1 \oplus g(W_o(\omega)) \oplus g(W_o(\bar{\omega})) \to g(W_{-1}(1)) \to 0$$

$$0 \to g(W_{1,\omega}(\bar{\omega})) \to P_{2_1} \oplus g(W_{2,\omega}(1)) \to g(W_{1,\omega}(\omega)) \to 0$$

$$0 \to g(W_{1,\bar{\omega}}(\omega)) \to P_{2_2} \oplus g(W_{2,\bar{\omega}}(1)) \to g(W_{1,\bar{\omega}}(\bar{\omega})) \to 0$$

The Auslander-Reiten quiver may thus be obtained from that for A_4 by relocating the projective modules as indicated above.

Cohomology

It again follows from the fact that a Sylow 2-subgroup of A_5 is a t.i. subgroup with normalizer A_4 (see 2.22 exercises 4 and 5) that

$$H^*(A_5,k) \cong H^*(A_4,k).$$

iii. Representations over $\mathbb{F}_3$ Representation type: finite

Decomposition matrix

	I	3_1	3_2	4
1	1			0
3_1		1		
3_2			1	
4	0			1
5	1			1

Cartan matrix

	I	4	3_1	3_2
I	2	1		
4	1	2		
3_1			1	
3_2				1

Atom Table and Representation Table for $A_k(G)$

	60	4	5	5	12	-4	-4	-4			6	2	12	-4	-4	-4
p power		A	A	A	A1A1	A1A2	A1A1	A1A2			A	A	A1A1	A1A2	A1B1	A1B2
ind	1A	2A	5A	B*	3A1	3A2	S3A1	S3A2	fus	ind	2B	4A	6A1	6A2	S3B1	S3B2
+	1	1	1	1	1	1	1	1	:	++	1	1	1	1	1	1
+	3	-1	-b5	*	0	0	0	0		+	0	0	0	0	0	0
+	3	-1	*	-b5	0	0	0	0								
+	4	0	-1	-1	1	1	-1	-1	:	++	-2	0	1	1	-1	-1
	0	0	0	0	3	-1	-1	-1	:		0	0	3	-1	-1	-1
	0	0	0	0	3	-1	1	1	:		0	0	3	-1	1	1
	0	0	0	0	-3	-1	i	-i	:		0	0	-3	-1	i	-i
	0	0	0	0	-3	-1	-i	i	:		0	0	-3	-1	-i	i

	60	4	5	5	12	-4	-4	-4		6	2	12	-4	-4	-4	
p power		A	A	A	A1A1	A1A2	A1A1	A1A2		A	A	A1A1	A1A2	A1B1	A1B2	
	1A	2A	5A	B*	3A1	3A2	S3A1	S3A2	fus	2B	4A	6A1	6A2	S3B1	S3B2	vtx
	6	2	1	1	0	0	0	0	:	0	2	0	0	0	0	1A
	3	-1	-b5	*	0	0	0	0		0	0	0	0	0	0	1A
	3	-1	*	-b5	0	0	0	0								1A
	9	1	-1	-1	0	0	0	0	:	-3	1	0	0	0	0	1A
	1	1	1	1	1	1	1	1	:	1	1	1	1	1	1	3A(1)
	4	0	-1	-1	1	1	-1	-1	:	-2	0	1	1	-1	-1	3A(1)
	5	1	0	0	-1	1	-i	i	:	-1	1	-1	1	-i	i	3A(2)
	5	1	0	0	-1	1	i	-i	:	-1	1	-1	1	i	-i	3A(2)

The Alternating Group A_6, and its coverings and automorphisms.

i. Ordinary Characters

									S_6							$PGL_2(9)$							M_{10}			
	360	8	9	9	4	5	5	.	.	24	24	4	3	3	.	.	10	4	4	5	5	.	.	2	4	4
p power		A	A	A	A	A	A			A	A	A	AB	BC			A	A	A	BD	AD			A	A	A
p' part		A	A	A	A	A	A			A	A	A	AB	BC			A	A	A	AD	BD			A	A	A
ind	1A	2A	3A	3B	4A	5A	B*	fus	ind	2B	2C	4B	6A	6B	fus	ind	2D	8A	B*	10A	B*	fus	ind	4C	8C	D**
+	1	1	1	1	1	1	1	:	++	1	1	1	1	1	:	++	1	1	1	1	1	:	++	1	1	1
+	5	1	2	-1	-1	0	0	:	++	3	-1	1	0	-1		+	0	0	0	0	0		+	0	0	0
+	5	1	-1	2	-1	0	0	:	++	1	-3	-1	1	0												
+	8	0	-1	-1	0	-b5	*		+	0	0	0	0	0	:	++	2	0	0	b5	*		+	0	0	0
+	8	0	-1	-1	0	*	-b5								:	++	2	0	0	*	b5					
+	9	1	0	0	1	-1	-1	:	++	3	3	-1	0	0	:	++	1	-1	-1	1	1	:	++	1	-1	-1
+	10	-2	1	1	0	0	0	:	++	2	-2	0	-1	1	:	++	0	r2	-r2	0	0	:	oo	0	i2	-i2
	1 2	4	3 6	3 6	8 8	5 10	5 10			2 2	4 4	8 8	6 6	12 12			4	16 16	16 16	20 20	20 20			4	16	16
-	4	0	-2	1	0	-1	-1	:	--	0	0	0	0	r3		-	0	0	0	0	0		-			
-	4	0	1	-2	0	-1	-1	:	oo	0	0	0	i3	0								*				
-	8	0	-1	-1	0	-b5	*		-	0	0	0	0	0	:	--	0	0	0	y20	*3		-			
-	8	0	-1	-1	0	*	-b5								:	--	0	0	0	*7	y20	*				
-	10	0	1	1	r2	0	0		-	0	0	0	0	0	:	--	0	y16	*5	0	0		-			
-	10	0	1	1	-r2	0	0								:	--	0	*13	y16	0	0	*				
	1 3 3	2 6 6	3	3	4 12 12	5 15 15	5 15 15			2	2	4	6	6			2	8	8	10	10			4 12 12	8 24 24	8 24 24
o2	3	-1	0	0	1	-b5	*		o2						*	+							o2	0	0	0
o2	3	-1	0	0	1	*	-b5	*							*	+										
o2	6	2	0	0	0	1	1	*	+						*	+						:	oo2	0	i2	-i2
o2	9	1	0	0	1	-1	-1	*	+						*	+						:	oo2	1	-1	-1
o2	15	-1	0	0	-1	0	0	*	+						*	+						:	oo2	1	1	1
	1 6 3 2 3 6	4 12 12	3 6	3 6	8 24 24 8 24 24	5 30 15 10 15 30	5 30 15 10 15 30			2	4	8	6 6	12 12			4	16 16	16 16	20 20	20 20			4 12 12	16 48 48	16 48 48
o2	6	0	0	0	r2	1	1		o2						*	-							o2			
o2	6	0	0	0	-r2	1	1	*							*	-						*7				
o2	12	0	0	0	0	b5	*		o2						*	-							o2			
o2	12	0	0	0	0	*	b5	*							*	-						*7				

ii. Representations over $\overline{\mathbb{F}}_2$ Representation type: tame

Decomposition Matrix

	I	4_1	4_2	8_1	8_2
1	1	0	0		
5	1	1	0		
5	1	0	1		
9	1	1	1		
10	2	1	1		
8				1	
8					1

Cartan Matrix

	I	4_1	4_2	8_1	8_2
I	8	4	4		
4_1	4	3	2		
4_2	4	2	3		
8_1				1	
8_2					1

Triple Cover

	3_1	3_2	9
3	1	0	0
3	0	1	0
6	1	1	0
9	0	0	1
15	1	1	1

	3_1	3_2	9
3_1	3	2	1
3_2	2	3	1
9	1	1	2

Atom Table and Representation Table for $A(A_6,Cyc)$ and $A(3A_6,Cyc)$ over $\overline{\mathbb{F}}_2$

atoms

	360	9	9	5	5	-8	-8	8
p power		A	A	A	A	A	A	A
ind	1A	3A	3B	5A	B*	2A	4A1	4A2
+	1	1	1	1	1	1	1	1
-	4	1	-2	-1	-1	0	0	0
-	4	-2	1	-1	-1	0	0	0
+	8	-1	-1	-b5	*	0	0	0
+	8	-1	-1	*	-b5	0	0	0
	0	0	0	0	0	-2	2	0
	0	0	0	0	0	0	-2	2
	0	0	0	0	0	0	-2	-2
	1	3	3	5	5	2	4	4
	3			15	15	6	12	12
	3			15	15	6	12	12
o2	3	0	0	-b5	*	1	1	-1
o2	3	0	0	*	-b5	1	1	-1
o2	9	0	0	-1	-1	1	1	1
	0	0	0	0	0	-2	2	0
	0	0	0	0	0	0	-2	2
	0	0	0	0	0	0	-2	-2

representations

	360	9	9	5	5	-8	-8	8
p power		A	A	A	A	A	A	A
	1A	3A	3A	5A	B*	2A	4A1	4A2
	40	4	4	0	0	0	0	0
	24	3	0	-1	-1	0	0	0
	24	0	3	-1	-1	0	0	0
	8	-1	-1	-b5	*	0	0	0
	8	-1	-1	*	-b5	0	0	0
	20	2	2	0	0	4	0	0
	10	1	1	0	0	2	2	2
	30	3	3	0	0	2	2	-2
	1	3	3	5	5	2	4	4
	3			15	15	6	12	12
	3			15	15	6	12	12
	24	0	0	*2+b5		0	0	0
	24	0	02+b5		*	0	0	0
	24	0	0	-1	-1	0	0	0
	36	0	0	1	1	4	0	0
	42	0	0	2	2	2	2	2
	30	0	0	0	0	2	2	-2

Atom Table and Representation Table for $A(2A_6,Cyc)$ over $\overline{\mathbb{F}}_2$

	720	18	18	10	10	-720	-18	-18	-10	-10	-16	16	-16	16	16	-16
p power		A	A	A	A	A	AA	BA	BA	AA	A	A	A	A	A	A
ind	1A	3A	3B	5A	B*	2A	6A	6B	10A	B*	4A1	4A2	8A1	8A2	8A3	8A4
+	1	1	1	1	1	1	1	1	1	1	1	1	1	1	1	1
-	4	1	-2	-1	-1	4	1	-2	-1	-1	0	0	0	0	0	0
-	4	-2	1	-1	-1	4	-2	1	-1	-1	0	0	0	0	0	0
+	8	-1	-1	-b5	*	8	-1	-1	-b5	*	0	0	0	0	0	0
+	8	-1	-1	*	-b5	8	-1	-1	*	-b5	0	0	0	0	0	0
	0	0	0	0	0	-2	-2	-2	-2	-2	2	0	2	0	0	-2
	0	0	0	0	0	-8	-2	4	2	2	0	0	0	0	0	0
	0	0	0	0	0	-8	4	-2	2	2	0	0	0	0	0	0
	0	0	0	0	0	-16	-2	2	2b5	*	0	0	0	0	0	0
	0	0	0	0	0	-16	-2	2	*	2b5	0	0	0	0	0	0
	0	0	0	0	0	0	0	0	0	0	-2	2	2	0	-2	0
	0	0	0	0	0	0	0	0	0	0	-2	-2	2	0	2	0
	0	0	0	0	0	0	0	0	0	0	0	0	-2	2	2	-2
	0	0	0	0	0	0	0	0	0	0	0	0	-2	-2	2	2
	0	0	0	0	0	0	0	0	0	0	0	0	-2	2	-2	2
	0	0	0	0	0	0	0	0	0	0	0	0	-2	-2	-2	-2

	720	18	18	10	10	-720	-18	-18	-10	-10	-16	16	-16	16	16	-16
p power		A	A	A	A	A	AA	BA	BA	AA	A	A	A	A	A	A
ind	1A	3A	3B	5A	B*	2A	6A	6B	10A	B*	4A1	4A2	8A1	8A2	8A3	8A4
	80	8	8	0	0	0	0	0	0	0	0	0	0	0	0	0
	48	6	0	-2	-2	0	0	0	0	0	0	0	0	0	0	0
	48	0	6	-2	-2	0	0	0	0	0	0	0	0	0	0	0
	16	-2	-2	-2b5	*	0	0	0	0	0	0	0	0	0	0	0
	16	-2	-2	*	-2b5	0	0	0	0	0	0	0	0	0	0	0
	40	4	4	0	0	40	4	4	0	0	0	0	2	0	0	0
	24	3	0	-1	-1	24	3	0	-1	-1	0	0	0	0	0	0
	24	0	3	-1	-1	24	0	3	-1	-1	0	0	0	0	0	0
	8	-1	-1	-b5	*	8	-1	-1	-b5	*	0	0	0	0	0	0
	8	-1	-1	*	-b5	8	-1	-1	*	-b5	0	0	0	0	0	0
	20	2	2	0	0	20	2	2	0	0	4	4	0	0	0	0
	60	6	6	0	0	20	2	2	0	0	4	-4	0	0	0	0
	10	1	1	0	0	10	1	1	0	0	2	2	2	2	2	2
	30	3	3	0	0	30	3	3	0	0	2	2	2	-2	2	-2
	50	5	5	0	0	30	3	3	0	0	2	-2	2	2	-2	-2
	70	7	7	0	0	10	1	1	0	0	2	-2	2	-2	-2	2

Projective Indecomposable Modules for A_6 over $\overline{\mathbb{F}}_2$

I		4_1	4_2		
4_1	4_2	I	I		
I	I	4_2	4_1		
4_2	4_1	I	I		
I	I	4_1	4_2	8_1	8_2
4_1	4_2	I	I		
I	I	4_2	4_1		
4_2	4_1	I	I		
I		4_1	4_2		

Projective Indecomposable Modules for $3A_6$ over $\overline{\mathbb{F}}_2$

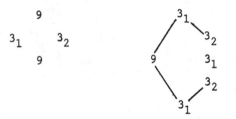

9

3_1 3_2

9

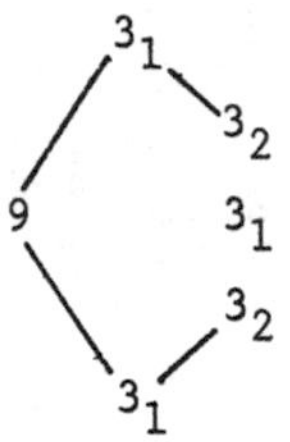

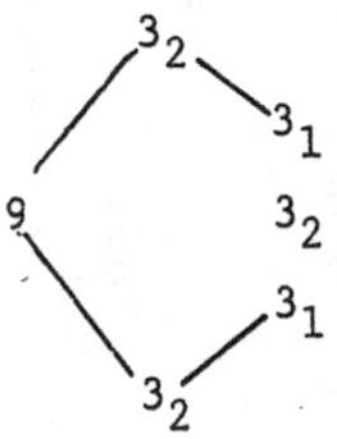

(and duals of these)

Projective Indecomposable modules for S_6 over $\mathbb{F}_2$

			I			
	4_1		I		4_2	
I		4_1		4_2		I
4_2		I		I		4_1
I		4_2		4_1		I
4_1		I		I		4_2
I		4_1		4_2		I
4_2		I		I		4_1
	4_2		I		4_1	
			I			

	4_1			4_2	
I		4_1	I		4_2
4_2		I	4_1		I
I		4_2	I		4_1
4_1		I	4_2		I
I		4_1	I		4_2
4_2		I	4_1		I
I		4_2	I		4_1
4_1		I	4_2		I
	4_1			4_2	

16

Projective indecomposable modules for M_{10} over $\mathbb{F}_2$

	I	
8		I
I		8
I		I
8		I
I		8
I		I
8		I
I		8
I		I
8		I
I		8
	I	

8
I
I
8
I
I
8
I
I
8
I
I
8

16

Projective indecomposable modules for $PGL_2(9)$ over $\mathbb{F}_2$

	I		8_1		
8_1		I	I		
I		8_1	I		
I		I	8_1		
8_1		I	I		
I		8_1	I		
I		I	8_1	8_2	8_3
8_1		I	I		
I		8_1	I		
I		I	8_1		
8_1		I	I		
I		8_1	I		
	I		8_1		

iii. Representations over $\mathbb{F}_3$

Representation type: wild

Decomposition Matrix

	I	3_1	3_2	4	9
1	1	0	0	0	
5	1	0	0	1	
5	1	0	0	1	
8	1	1	0	1	
8	1	0	1	1	
10	0	1	1	1	
9					1

Cartan Matrix

	I	3_1	3_2	4	9
I	5	1	1	4	
3_1	1	2	1	2	
3_2	1	1	2	2	
4	4	2	2	5	
9					1

Double Cover

	2_1	2_2	6_1	6_2
4	1	1	0	0
4	1	1	0	0
8	0	1	1	0
8	1	0	0	1
10	1	1	1	0
10	1	1	0	1

	2_1	2_2	6_1	6_2
2_1	5	4	1	2
2_2	4	5	2	1
6_1	1	2	2	0
6_2	2	1	0	2

Projective Indecomposable Modules for A_6 over $\bar{\mathbb{F}}_3$

$$\begin{matrix} & & I & & \\ & 4 & & 4 & \\ I & I & I & 3_1 & 3_2 \\ & 4 & & 4 & \\ & & I & & \end{matrix} \qquad \begin{matrix} & 3_1 & \\ & 4 & \\ I & & 3_2 \\ & 4 & \\ & 3_1 & \end{matrix} \qquad \begin{matrix} & 3_2 & \\ & 4 & \\ I & & 3_1 \\ & 4 & \\ & 3_2 & \end{matrix}$$

$$\begin{matrix} & & 4 & & \\ I & I & & 3_1 & 3_2 \\ & 4 & 4 & 4 & \\ I & I & & 3_1 & 3_2 \\ & & 4 & & \end{matrix} \qquad 9$$

Double Cover

$$\begin{matrix} & & 2_1 & & \\ & 2_2 & 2_2 & 6_2 & \\ 2_1 & 2_1 & 2_1 & 6_1 & \\ & 2_2 & 2_2 & 6_2 & \\ & & 2_1 & & \end{matrix} \qquad \begin{matrix} & & 2_2 & & \\ & 2_1 & 2_1 & 6_1 & \\ 2_2 & 2_2 & 2_2 & 6_2 & \\ & 2_1 & 2_1 & 6_1 & \\ & & 2_2 & & \end{matrix} \qquad \begin{matrix} 6_1 \\ 2_2 \\ 2_1 \\ 2_2 \\ 6_1 \end{matrix} \qquad \begin{matrix} 6_2 \\ 2_1 \\ 2_2 \\ 2_1 \\ 6_2 \end{matrix}$$

iv. Representations over $\bar{\mathbb{F}}_5$: Brauer character table

Representation type: finite

	360	8	9	9	4			24	24	4	3	3			10	4	4			2	4	4
p power		A	A	A	A			A	A	A	AB	BC			A	A	A			A	A	A
p' part		A	A	A	A			A	A	A	AB	BC			A	A	A			A	A	A
	1A	2A	3A	3B	4A	fus	ind	2B	2C	4B	6A	6B	fus	ind	2D	8A	B*	fus	ind	4C	8C	D**
+	1	1	1	1	1	:	++	1	1	1	1	1	:	++	1	1	1	:	++	1	1	1
+	5	1	2	-1	-1	:	++	3	-1	1	0	-1	┬	+	0	0	0	┬	+	0	0	0
+	5	1	-1	2	-1	:	++	-1	3	1	-1	0	┴					┴				
+	8	0	-1	-1	0	:	++	2	2	-2	-1	-1	:	++	2	0	0	:	++	0	-2	-2
+	10	-2	1	1	0	:	++	2	-2	0	-1	1	:	++	0	r2	-r2	:	oo	0	i2	-i2
ind	1	4	3	3	8	fus	ind	2	4	8	6	12	fus	ind	4	16	16	fus	ind	4	16	16
	2		6	6	8						6	12				16	16					
-	4	0	-2	1	0	:	--	0	0	0	0	r3	┬	-	0	0	0	┬	o			
-	4	0	1	-2	0	:	oo	0	0	0	i3	0	┴					*				
-	10	0	1	1	r2	┬	-	0	0	0	0	0	:	--	0	y16	*5	┬	o			
-	10	0	1	1	-r2	┴							:	--	0	*13	y16	*				
ind	1	2	3	3	4	fus	ind	2	2	4	6	6	fus	ind	2	8	8	fus	ind	4	8	8
	3	6			12															12	24	24
	3	6			12															12	24	24
o2	3	-1	0	0	1	*	+						*	+				:	oo2	1	i2-1	**
o2	6	2	0	0	0	*	+						*	+				:	oo2	0	i2	-i2
o2	15	-1	0	0	-1	*	+						*	+				:	oo2	1	1	1
ind	1	4	3	3	8	fus	ind	2	4	8	6	12	fus	ind	4	16	16	fus	ind	4	16	16
	6	12	6	6	24						6	12				16	16			12	48	48
	3	12			24															12	48	48
	2				8																	
	3				24																	
	6				24																	
o2	6	0	0	0	r2	┬	o2						*	-				┬	o2			
o2	6	0	0	0	-r2	*							*	-				*7				

The Alternating Group A_7, and its coverings and automorphisms.

i. Ordinary Characters

	2520	24	36	9	4	5	12	7	7			120	24	12	6	3	5	6
p power		A	A	A	A	A	A	A	A			A	A	A	AB	BC	AB	AB
p' part		A	A	A	A	A	A	A	A			A	A	A	AB	BC	AB	AB
ind	1A	2A	3A	3B	4A	5A	6A	7A	B**	fus	ind	2B	2C	4B	6B	6B	10A	12A
+	1	1	1	1	1	1	1	1	1	:	++	1	1	1	1	1	1	1
+	6	2	3	0	0	1	-1	-1	-1	:	++	4	0	2	1	0	-1	-1
o	10	-2	1	1	0	0	1	b7	**	•	+	0	0	0	0	0	0	0
o	10	-2	1	1	0	0	1	**	b7	•								
+	14	2	2	-1	0	-1	2	0	0	:	++	6	2	0	0	-1	1	0
+	14	2	-1	2	0	-1	-1	0	0	:	++	4	0	-2	1	0	-1	1
+	15	-1	3	0	-1	0	-1	1	1	:	++	5	-3	1	-1	0	0	1
+	21	1	-3	0	-1	1	1	0	0	:	++	1	-3	-1	1	0	1	-1
+	35	-1	-1	-1	1	0	-1	0	0	:	++	5	1	-1	-1	1	0	-1
	1	4	3	3	8	5	12	7	7			2	4	8	6	12	10	24
	2		6	6	8	10		14	14							12	10	24
o	4	0	-2	1	0	-1	0	-b7	**	•	-	0	0	0	0	0	0	0
o	4	0	-2	1	0	-1	0	**	-b7	•								
-	14	0	2	-1	r2	-1	0	0	0	•	-	0	0	0	0	0	0	0
-	14	0	2	-1	-r2	-1	0	0	0	•								
-	20	0	-4	-1	0	0	0	-1	-1	:	--	0	0	0	0	r3	0	0
-	20	0	2	2	0	0	0	-1	-1	:	--	0	0	0	0	0	0	r6
-	36	0	0	0	0	1	0	1	1	:	--	0	0	0	0	0	r5	0
	1	2	3	3	4	5	6	7	7			2	2	4	6	6	10	12
	3	6			12	15	6	21	21									
	3	6			12	15	6	21	21									
o2	6	2	0	0	0	1	2	-1	-1	*	+							
o2	15	-1	0	0	-1	0	2	1	1	*	+							
o2	15	3	0	0	1	0	0	1	1	*	+							
o2	21	1	0	0	-1	1	-2	0	0	*	+							
o2	21	-3	0	0	1	1	0	0	0	*	+							
o2	24	0	0	0	0	-1	0	b7	**	•	+2							
o2	24	0	0	0	0	-1	0	**	b7	*								
	1	4	3	3	8	5	12	7	7			2	4	8	6	12	10	24
	6	12	6	6	24	30	12	42	42							12	10	24
	3	12			24	15	12	21	21									
	2				8	10		14	14									
	3				24	15		21	21									
	6				24	30		42	42									
o2	6	0	0	0	r2	1	0	-1	-1	•	o2							
o2	6	0	0	0	-r2	1	0	-1	-1	*	-2							
o2	24	0	0	0	0	-1	0	b7	**	•								
o2	24	0	0	0	0	-1	0	**	b7	*								
o2	36	0	0	0	0	1	0	1	1	*	-							

ii. Representations over $\mathbb{F}_2$

Representation type: tame

Decomposition Matrix

	I	14	20	4_1	4_2	6
1	1	0	0			
15	1	1	0			
21	1	0	1			
35	1	1	1			
14	0	1	0			
6				0	0	1
10				0	1	1
10				1	0	1
14				1	1	1

Cartan Matrix

	I	14	20	4_1	4_2	6
I	4	2	3			
14	2	3	1			
20	2	1	2			
4_1				2	1	2
4_2				1	2	2
6				2	2	4

Triple Cover

	6	15	24_1	24_2
6	1	0		
15	0	1		
15	0	1		
21	1	1		
21	1	1		
24			1	
24				1

	6	15	24_1	24_2
6	3	2		
15	2	4		
24_1			1	
24_2				1

Atom Table and Representation Table for $A(A_7,\mathrm{Cyc})$ over $\bar{\mathbb{F}}_2$

	2520	36	9	5	7	7	-24	-8	8	-12
p power		A	A	A	A	A	A	A	A	AA
ind	1A	3A	3B	5A	7A	B**	2A	4A1	4A2	6A
+	1	1	1	1	1	1	1	1	1	1
o	4	-2	1	-1	-b7	**	0	0	0	0
o	4	-2	1	-1	**	-b7	0	0	0	0
+	6	3	0	1	-1	-1	2	0	0	-1
+	14	2	-1	-1	0	0	2	0	0	2
-	20	-4	-1	0	-1	-1	0	0	0	0
	0	0	0	0	0	0	-2	2	0	-2
	0	0	0	0	0	0	-4	0	0	2
	0	0	0	0	0	0	0	-2	2	0
	0	0	0	0	0	0	0	-2	-2	0

2520	36	9	5	7	7	-24	-8	8	-12
p power	A	A	A	A	A	A	A	A	AA
1A	3A	3B	5A	7A	B**	2A	4A1	4A2	6A
72	0	0	2	2	2	0	0	0	0
24	0	3	-1	b7	**	0	0	0	0
24	0	3	-1	**	b7	0	0	0	0
40	4	4	0	-2	-2	0	0	0	0
64	4	-2	-1	1	1	0	0	0	0
56	-4	-1	1	0	0	0	0	0	0
100	4	-2	0	2	2	4	0	0	4
20	2	2	0	-1	-1	4	0	0	-2
50	2	-1	0	1	1	2	2	2	2
86	2	-1	1	2	2	2	2	-2	2

Projective Indecomposable Modules for A_7 over $\bar{\mathbb{F}}_2$

I
14 20
I I
20 14
I

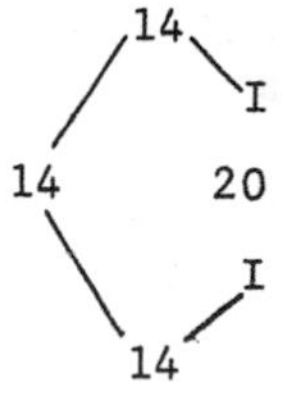

20 4_1 4_2
I 6 6
14 4_2 4_1
I 6 6
20 4_1 4_2

6
4_1 4_2
6 6
4_2 4_1
6

iii. Representations over $\bar{\mathbb{F}}_3$

Representation type: wild

Decomposition Matrix

	I	10_1	10_2	13	6	15
1	1	0	0	0		
10	0	1	0	0		
10	0	0	1	0		
14	1	0	0	1		
14	1	0	0	1		
35	2	1	1	1		
6					1	0
15					0	1
21					1	1

Cartan Matrix

	I	10_1	10_2	13	6	15
I	7	2	2	4		
10_1	2	2	1	1		
10_2	2	1	2	1		
13	4	1	1	3		
6					2	1
15					1	2

Projective Indecomposable Modules for A_7 over $\bar{\mathbb{F}}_3$

I
10_1 10_2 13 13
I I I I I
10_1 10_2 13 13
I

10_1
I
10_2 13
I
10_1

10_2
I
10_1 13
I
10_2

13
I I
10_1 10_2 13
I I
13

6
15
6

15
6
15

The Linear Group $L_3(2)$, and its coverings and automorphisms.

i. Ordinary Characters

	168	8	3	4	7	7			6	3	4	4
p power		A	A	A	A	A			A	AB	A	A
p' part		A	A	A	A	A			A	AB	A	A
ind	1A	2A	3A	4A	7A	B**	fus	ind	2B	6A	8A	B*
+	1	1	1	1	1	1	:	++	1	1	1	1
o	3	-1	0	1	b7	**		+	0	0	0	0
o	3	-1	0	1	**	b7						
+	6	2	0	0	-1	-1	:	++	0	0	r2	-r2
+	7	-1	1	-1	0	0	:	++	1	1	-1	-1
+	8	0	-1	0	1	1	:	++	2	-1	0	0
ind	1 2	4	3 6	8 8	7 14	7 14	fus	ind	4	12 12	16 16	16 16
o	4	0	1	0	-b7	**		-	0	0	0	0
o	4	0	1	0	**	-b7						
-	6	0	0	r2	-1	-1	:	--	0	0	y16	*3
-	6	0	0	-r2	-1	-1	:	--	0	0	*5	y16
-	8	0	-1	0	1	1	:	--	0	r3	0	0

ii. Representations over $\mathbb{F}_2$

Representation type: tame

Decomposition Matrix

	I	3_1	3_2	8
1	1	0	0	
3	0	1	0	
3	0	0	1	
6	0	1	1	
7	1	1	1	
8				1

Cartan Matrix

	I	3_1	3_2	8
I	2	1	1	
3_1	1	3	2	
3_2	1	2	3	
8				1

Atom Table and Representation Table for $A(L_3(2),Cyc)$ over $\bar{\mathbb{F}}_2$

	168	3	7	7	-8	-8	8
p power		A	A	A	A	A	A
ind	1A	3A	7A	B**	2A	4A1	4A2
+	1	1	1	1	1	1	1
o	3	0	b7	**	1	1	-1
o	3	0	**	b7	1	1	-1
+	8	-1	1	1	0	0	0
	0	0	0	0	-2	2	0
	0	0	0	0	0	-2	2
	0	0	0	0	0	-2	-2

	168	3	7	7	-8	-8	8	
p power		A	A	A	A	A	A	
	1A	3A	7A	B**	2A	4A1	4A2	vtx
	8	2	1	1	0	0	0	1A
	16	1	b7-1	**	0	0	0	1A
	16	1	**	b7-1	0	0	0	1A
	8	-1	1	1	0	0	0	1A
	20	2	-1	-1	4	0	0	2A
	26	2	-2	-2	2	2	2	4A(1)
	14	2	0	0	2	2	-2	4A(3)

Projective Indecomposable Modules for $L_3(2)$ over $\bar{\mathbb{F}}_2$

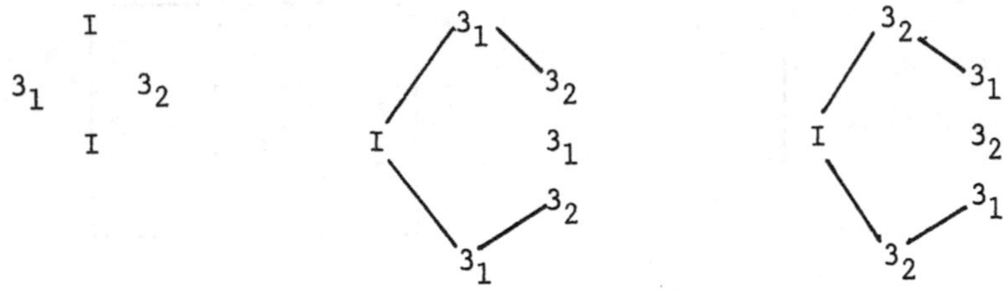

8

Atom Table and Representation Table for $A(2^3{:}L_3(2),Cyc)$ over $\mathbb{F}_2$

	1A	3A	7A	B**	2A	2B	2C	6A	4A1	4A2	4B1	4B2	4C1	4C2
	1344	6	7	7	-192	-32	-32	-6	-16	16	-8	8	-8	8
p power		A	A	A	A	A	A	AA	A	A	B	B	C	C
ind	1A	3A	7A	B**	2A	2B	2C	6A	4A1	4A2	4B1	4B2	4C1	4C2
+	1	1	1	1	1	1	1	1	1	1	1	1	1	1
o	3	0	b7	**	3	1	1	0	1	1	1	-1	1	-1
o	3	0	**	b7	3	1	1	0	1	1	1	-1	1	-1
+	8	-1	1	1	8	0	0	-1	0	0	0	0	0	0
	0	0	0	0	0	-2	0	0	0	0	2	0	0	0
	0	0	0	0	0	0	0	0	0	0	-2	2	0	0
	0	0	0	0	0	0	0	0	0	0	-2	-2	0	0
	0	0	0	0	0	0	-2	0	0	0	0	0	2	0
	0	0	0	0	0	0	0	0	0	0	0	0	-2	2
	0	0	0	0	0	0	0	0	0	0	0	0	-2	-2
	0	0	0	0	-2	0	0	-2	2	0	0	0	0	0
	0	0	0	0	0	0	0	0	-2	2	0	0	0	0
	0	0	0	0	0	0	0	0	-2	-2	0	0	0	0
	0	0	0	0	-4	0	0	2	0	0	0	0	0	0

	1A	3A	7A	B**	2A	2B	2C	6A	4A1	4A2	4B1	4B2	4C1	4C2	
	1344	6	7	7	-192	-32	-32	-6	-16	16	-8	8	-8	8	
p power		A	A	A	A	A	A	AA	A	A	B	B	C	C	
	1A	3A	7A	B**	2A	2B	2C	6A	4A1	4A2	4B1	4B2	4C1	4C2	vtx
	64	4	1	1	0	0	0	0	0	0	0	0	0	0	1A
	128	2	b7-1	**	0	0	0	0	0	0	0	0	0	0	1A
	128	2	**	b7-1	0	0	0	0	0	0	0	0	0	0	1A
	64	-2	1	1	0	0	0	0	0	0	0	0	0	0	1A
	160	4	-1	-1	0	16	0	0	0	0	0	0	0	0	2B
	208	4	-2	-2	0	8	0	0	0	0	2	2	0	0	4B(1)
	112	4	0	0	0	8	0	0	0	0	2	-2	0	0	4B(3)
	160	4	-1	-1	0	0	16	0	0	0	0	0	0	0	2C
	208	4	-2	-2	0	0	8	0	0	0	0	0	2	2	4C(1)
	112	4	0	0	0	0	8	0	0	0	0	0	2	-2	4C(3)
	160	4	-1	-1	96	0	0	0	0	0	0	0	0	0	2A
	208	4	-2	-2	48	0	0	0	4	4	0	0	0	0	4A(1)
	112	4	0	0	48	0	0	0	4	-4	0	0	0	0	4A(3)
	224	-1	0	0	32	0	0	-1	0	0	0	0	0	0	2A

The Sporadic Group M_{11}

i. Ordinary Characters

7920	48	18	8	5	6	8	8	11	11
p power	A	A	A	A	AA	A	A	A	A
p' part	A	A	A	A	AA	A	A	A	A
ind 1A	2A	3A	4A	5A	6A	8A	B**	11A	B**
+ 1	1	1	1	1	1	1	1	1	1
+ 10	2	1	2	0	-1	0	0	-1	-1
o 10	-2	1	0	0	1	i2	-i2	-1	-1
o 10	-2	1	0	0	1	-i2	i2	-1	-1
+ 11	3	2	-1	1	0	-1	-1	0	0
o 16	0	-2	0	1	0	0	0	b11	**
o 16	0	-2	0	1	0	0	0	**	b11
+ 44	4	-1	0	-1	1	0	0	0	0
+ 45	-3	0	1	0	0	-1	-1	1	1
+ 55	-1	1	-1	0	-1	1	1	0	0

ii. Representations over $\bar{\mathbb{F}}_2$

Representation type: tame

Decomposition Matrix

	I	10	44	16_1	16_2
1	1	0	0		
10	0	1	0		
10	0	1	0		
10	0	1	0		
11	1	1	0		
44	0	0	1		
45	1	0	1		
55	1	1	1		
16				1	
16					1

Cartan Matrix

	I	10	44	16_1	16_2
I	4	2	2		
10	2	5	1		
44	2	1	3		
16_1				1	
16_2					1

Atom Table and Representation Table for $A(M_{11},Cyc)$ over $\bar{\mathbb{F}}_2$

	7920	18	5	11	11	-48	-6	-16	16	-16	16	16	-16
p power		A	A	A	A	A	AA	A1	A2	A1	A2	A3	A4
ind	1A	3A	5A	11A	B**	2A	6A	4A1	4A2	8A1	8A2	8A3	8A4
+	1	1	1	1	1	1	1	1	1	1	1	1	1
+	10	1	0	-1	-1	2	-1	2	2	0	0	0	0
o	16	-2	1	b11	**	0	0	0	0	0	0	0	0
o	16	-2	1	**	b11	0	0	0	0	0	0	0	0
+	44	-1	-1	0	0	4	1	0	0	0	0	0	0
	0	0	0	0	0	-2	-2	2	0	2	0	0	-2
	0	0	0	0	0	-4	2	4	0	-2	-2	2	2
	0	0	0	0	0	0	0	-2	2	2	0	-2	0
	0	0	0	0	0	0	0	-2	-2	2	0	2	0
	0	0	0	0	0	0	0	0	0	-2	2	2	-2
	0	0	0	0	0	0	0	0	0	-2	-2	2	2
	0	0	0	0	0	0	0	0	0	-2	2	-2	2
	0	0	0	0	0	0	0	0	0	-2	-2	-2	-2

7920	18	5	11	11	-48	-6	-16	16	-16	16	16	-16	
p power	A	A	A	A	A	AA	A1	A2	A1	A2	A3	A4	
1A	3A	5A	11A	B**	2A	6A	4A1	4A2	8A1	8A2	8A3	8A4	vtx
112	4	2	2	0	0	0	0	0	0	0	0	0	1A
96	[illegible]	1	-3	-3	0	0	0	0	0	0	0	0	1A
16	-2	1	b11	**	0	0	0	0	0	0	0	0	1A
16	-2	1	**	b11	0	0	0	0	0	0	0	0	1A
144	0	1	1	1	0	0	0	0	0	0	0	0	1A
200	2	0	2	2	8	2	0	0	0	0	0	0	2A
120	3	0	-1	-1	8	-1	0	0	0	0	0	0	2A
220	4	0	0	0	12	0	4	4	0	0	0	0	4A(1)
372	12	2	-2	-2	12	0	4	-4	0	0	0	0	4A(3)
110	2	0	0	0	6	0	2	2	2	2	2	2	8A(1)
90	0	0	2	2	2	2	2	2	2	-2	2	-2	8A(3)
286	7	1	0	0	10	1	2	-2	2	2	-2	-2	8A(5)
242	8	2	0	0	6	0	2	-2	2	-2	-2	2	8A(7)

Projective Indecomposable Modules for M_{11} over $\bar{\mathbb{F}}_2$

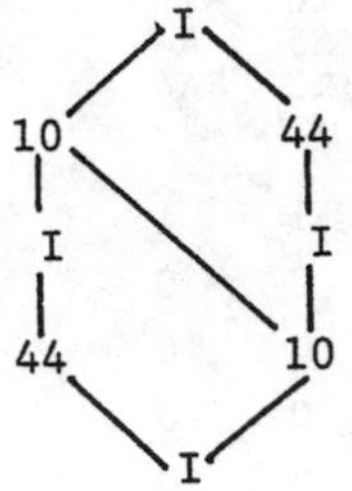

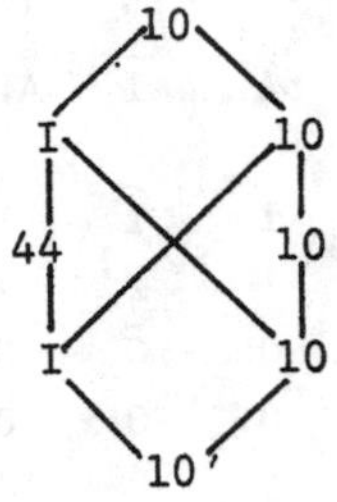

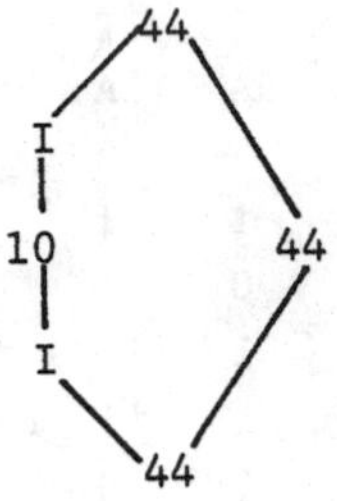

16_1 , 16_2

References

[1] J. L. Alperin, 'Periodicity in Groups', Ill. J. Math. 21 (1977) 776-83.

[2] J. L. Alperin, L. Evens, 'Representations, resolutions, and Quillen's dimension theorem', J. Pure Appl. Alg. 22, 1-9 (1981).

[3] J. L. Alperin, L. Evens, 'Varieties and elementary abelian subgroups', J. Pure Appl. Alg. 26 (1982) 221-227.

[4] M. F. Atiyah, I. G. Macdonald, 'Introduction to Commutative Algebra', Addison-Wesley, Reading, Mass. (1969).

[5] M. Auslander, I. Reiten, 'Representation Theory of Artin Algebras, III: almost split sequences', Comm. in Alg. 3 (3), 239-294 (1975).

[6] M. Auslander, 'Existence theorems for almost split sequences', Oklahoma Ring Theory Conference, March 1976.

[7] M. Auslander, I. Reiten, 'Representation theory of Artin Algebras, IV: invariants given by almost split sequences', Comm. in Alg. 5 (5), 443-518 (1977).

[8] G. S. Avrunin, 'Annihilators of cohomology modules', J. Alg. 69, 150-154 (1981).

[9] G. S. Avrunin and L. L. Scott, 'Quillen Stratification for Modules', Invent. Math. 66, 277-286 (1982).

[10] V. A. Bašev, 'Representations of the group $Z_2 \times Z_2$ in a field of characteristic 2 ', (Russian), Dokl. Akad. Nauk . SSSR 141 (1961), 1015-1018.

[11] D. J. Benson, 'The Loewy structure of the projective indecomposable modules for A_8 in characteristic two', Commun. in Alg., 11 (13), 1395-1432, 1983.

[12] D. J. Benson, 'The Loewy structure of the projective indecomposable modules for A_9 in characteristic two', Commun. in Alg., 11 (13), 1433-1453, 1983.

[13] D. J. Benson. R. A. Parker, 'The Green Ring of a Finite Group', J. Alg. 87, 290-331 (1984).

[14] D. J. Benson, 'Lambda and Psi Operations on Green Rings', J. Alg. 87, 360-367 (1984).

[15] Berman, Moody and Wonenburger, 'Cartan Matrices with Null Roots and Finite Cartan Matrices', Indiana U. Math. J. 21 (7-12) 1972, 1091.

[16] R. Brauer, Collected Papers, Vols. I-III, MIT Press, 1980, ed. P. Fong and W. Wong.

[17] D. Burry, J. Carlson, 'Restrictions of modules to local subgroups', Proc. A.M.S. 84, 181-184, 1982.

[18] J. F. Carlson, 'The modular representation ring of a cyclic 2-group', J. L.M.S. (2), 11 (1975), 91-92.

[19] J. F. Carlson, 'Periodic Modules over Group Algebras', J. L.M.S. (2), 15 (1977), 431-436.

[20] J. F. Carlson, 'Restrictions of modules over modular group algebras', J. Alg. 53 (1978), 334-343.

[21] J. F. Carlson, 'The dimensions of periodic modules over modular group algebras', Ill. J. Math. 23 (1979), 295-306.

[22] J. F. Carlson, 'The complexity and varieties of modules', in 'Integral Representations and their applications', Lecture Notes in Mathematics 882, p. 415-422, Springer-Verlag 1981.

[23] J. F. Carlson, 'Complexity and Krull Dimension', in 'Representations of algebras', Lecture Notes in Mathematics 903, p. 62-67, Springer-Verlag 1981.

[24] J. F. Carlson, 'The structure of periodic modules over modular group algebras', J. Pure Appl. Alg. 22 (1981), 43-56.

[25] J. F. Carlson, 'Dimensions of modules and their restrictions over modular group algebras', J. Alg. 69, 95-104 (1981).

[26] J. F. Carlson, 'Varieties and the cohomology ring of a module', J. Alg. 85 (1983), 104-143.

[27] J. Carlson, 'The variety of an indecomposable module is connected', to appear, Invent. Math. 1984.

[28] R. Carter, 'Simple groups of Lie type', Wiley-Interscience, 1972.

[29] L. Chouinard, 'Projectivity and relative projectivity over group rings', J. Pure Appl. Alg. 7 (1976) 278-302.

[30] S. B. Conlon, 'Twisted Group algebras and their representations', J. Aust. Math. Soc., 4 (1964) 152-173.

[31] S. B. Conlon, 'Certain representation algebras', J. Aust. Math. Soc., 5 (1965) 83-99.

[32] S. B. Conlon, 'The modular representation algebra of groups with Sylow 2-subgroups $Z_2 \times Z_2$ ', J. Aust. Math. Soc. 6 (1966), 76-88.

[33] S. B. Conlon, 'Structure in representation algebras', J. Alg. 5 (1967), 274-279.

[34] S. B. Conlon, 'Relative components of representations', J. Alg. 8 (1968), 478-501.

[35] S. B. Conlon, 'Decompositions induced from the Burnside algebra', J. Alg. 10 (1968), 102-122.

[36] J. Conway, R. Curtis, S. Norton, R. Parker and R. Wilson, 'An Atlas of Finite Groups', to appear, OUP 1985.

[37] C. W. Curtis and I. Reiner, 'Representation theory of finite groups and associative algebras', Wiley-Interscience 1962.

[38] C. W. Curtis and I. Reiner, 'Methods in representation theory', Vol. I, J. Wiley and Sons, 1981.

[39] E. C. Dade, 'Endo-permutation modules over p-groups II', Ann. of Math. 108 (1978), 317-346.

[40] L. E. Dickson, 'On the algebra defined for any given field by the multiplication table of any given finite group', Trans. AMS 3, 285-301, 1902.

[41] V. Dlab and C. M. Ringel, 'Indecomposable representations of graphs and algebras', Memoirs of the A.M.S. (6) 173, 1976.

[42] P. Donovan and M. R. Freislich, 'Representable functions on the category of modular representations of a finite group with cyclic Sylow subgroups', J. Alg. 32, 356-364 (1974).

[43] P. Donovan and M. R. Freislich, 'Representable functions on the category of modular representations of a finite group with Sylow subgroup $C_2 \times C_2$ ', J. Alg. 32, 365-369 (1974).

[44] L. Dornhoff, 'Group representation theory, part B', Marcel Dekker, New York, 1972.

[45] A. Dress, 'On relative Grothendieck rings', Repn. Thy., Proc. Ottawa Conf. SLN 488, Springer, Berlin 1975.

[46] A. Dress, 'Modules with trivial source, modular monomial representations and a modular version of Brauer's induction theorem', Abh. Math. Sem. Univ. Hamburg 44 (1975), 101-109.

[47] D. Eisenbud, 'Homological algebra on a complete intersection, with an application to group representations', Trans. A.M.S. 260 (1980) 35-64.

[48] L. Evens, 'The Cohomology ring of a finite group', Trans. A.M.S. 101, 224-239 (1961).

[49] L. Evens, 'A generalization of the transfer map in the cohomology of groups', Trans. A.M.S. 108 (1963) 54-65.

[50] L. Evens, 'The Spectral Sequence of a Finite group extension stops', Trans. A.M.S. 212, 269-277 (1975).

[51] W. Feit, 'The representation theory of finite groups', North Holland, 1982.

[52] P. Gabriel, C. Riedtmann, 'Group representations without groups', Comm. Math. Helvetici 54 (1979) 240-287.

[53] P. Gabriel, 'Auslander-Reiten sequences and representation-finite algebras', Repn. Thy. I, Proc. Ottawa Conf. SLN 831, Springer Berlin 1980.

[54] J. A. Green, 'On the Indecomposable Representations of a finite group', Math. Zeit., Bd. 70 (1959), S. 430-445.

[55] J. A. Green, 'The modular representation algebra of a finite group', Ill. J. Math. 6 (4) (1962), 607-619.

[56] J. A. Green, 'Some remarks on defect groups', Math. Z. 107, 133-150, 1968.

[57] J. A. Green, 'A transfer theorem for modular representations', Trans. Amer. Math. Soc. 17 (1974), 197-213.

[58] D. Happel, U. Preiser, C. M. Ringel, 'Vinberg's characterization of Dynkin diagrams using subadditive functions with applications to DTr-periodic modules', Repn. Thy. II, Proc. Ottawa Conf. SLN 832, Springer, Berlin 1980.

[59] D. G. Higman, 'Indecomposable representations at characteristic p ', Duke Math. J. 21, 377-381, 1954.

[60] G. Hochschild, J.-P. Serre, 'Cohomology of group extensions', Trans. A.M.S. 74 (1953), 110-134.

[61] S. A. Jennings, 'The structure of the group ring of a p-group over a Modular Field', Trans. A.M.S. 50 (1941), 175-185.

[62] D. Knutson, ' λ-rings and the representation theory of the symmetric group', SLN 308, Springer, Berlin 1973.

[63] O. Kroll, 'Complexity and elementary abelian subgroups', Ph.D. thesis, Univ. of Chicago, 1980.

[64] P. Landrock and G. O. Michler, 'Block structure of the smallest Janko group', Math. Ann. 232, 205-238, 1978.

[65] P. Landrock, 'Finite group algebras and their modules', L.M.S. lecture note series, 1984.

[66] G. Lewis, 'The Integral Cohomology Rings of Groups of Order p^3', Trans. A.M.S. 132, 501-529, 1968.

[67] S. MacLane, 'Homology', Springer-Verlag, 1974.

[68] J. Milnor, 'The Steenrod algebra and its dual', Ann. of Math. vol. 67, no. 1, 150-171, 1958.

[69] H. J. Munkholm, 'Mod 2 cohomology of $D2^n$ and its extensions by Z_2 ', Conference on Alg. Topology, Univ. of Illinois at Chicago Circle, June 17-28, 1968, p. 234-252.

[70] H. Nagao, 'A proof of Brauer's Theorem on generalized decomposition numbers', Nagao Math. J. 22, 73-77, 1963.

[71] D. Quillen, 'A Cohomological criterion for p-nilpotence', J. Pure Appl. Alg. 1, 361-372 (1971).

[72] D. Quillen, 'The spectrum of an equivariant cohomology ring, I ', Ann. of Math. 94, 549-572 (1971).

[73] D. Quillen, 'The spectrum of an equivariant cohomology ring, II ', Ann. of Math. 94, 573-602 (1971).

[74] D. Quillen, 'The Mod 2 Cohomology Rings of Extra-Special 2-groups and the Spinor Groups', Math. Ann. 194, 197-212 (1971).

[75] D. Quillen, B. B. Venkov, 'Cohomology of finite groups and elementary abelian subgroups', Topology 11, 317-318 (1972).

[76] I. Reiten, 'Almost Split Sequences', Workshop on permutation groups and indecomposable modules, Giessen, September, 1975.

[77] I. Reiten, 'Almost split sequences for group algebras of finite representation type', Trans. A.M.S. 335, 125-136, 1977.

[78] C. Riedtmann, 'Algebren, Darstellungsköcher, Ueberlagerungen und Zurück', Comm. Math. Helvetici 55 (1980), 199-224.

[79] C. M. Ringel, 'The Indecomposable Representations of the Dihedral 2-groups', Math. Ann. 214, 19-34, 1975.

[80] K. W. Roggenkamp and J. W. Schmidt, 'Almost split sequences for integral group rings and orders', Commun. in Alg. 4, 893-917, 1976.

[81] K. W. Roggenkamp, 'The construction of almost split sequences for integral group rings and orders', Commun. in Alg., 5 (13), 1363-1373, 1977.

[82] J. Sawka, 'Odd primary operations in first-quadrant spectral sequences', Trans. A.M.S. 273 (2), 1982, 737-752.

[83] L. L. Scott, 'Modular permutation representations', Trans. A.M.S. 175, 101-121, 1973.

[84] J. P. Serre, 'Homologie Singulière des Espaces Fibrés', Annals of Maths. 54, 3 (1951), 425-505.

[85] J. P. Serre, 'Sur la dimension cohomologique des groupes profinis', Topology 3, 413-420 (1965).

[86] W. M. Singer, 'Steenrod Squares in Spectral Sequences, I ', Trans. A.M.S. 175, 1973, 327-336.

[87] W. M. Singer, 'Steenrod Squares in Spectral Sequences, II ', Trans. A.M.S. 175, 1973, 337-353.

[88] W. Smoke, 'Dimension and Multiplicity for Graded Algebras', J. Alg. 21, 149-173 (1972).

[89] N. Steenrod, 'Cohomology operations, and obstructions to extending continuous functions', Colloquium Lectures, 1957.

[90] N. Steenrod, 'Cohomology operations', Ann. of Math. Studies no. 50, Princeton 1962 , (Notes by D. Epstein).

[91] R. G. Swan, 'Induced representations and projective modules', Ann. of Math. vol. 71, no. 3, 552-578, 1960.

[92] P. J. Webb, 'The Auslander-Reiten quiver of a finite group', Math. Z. 179, 97-121, 1982.

[93] P. J. Webb, 'On the Orthogonality Coefficients for Character Tables of the Green Ring of a Finite Group', preprint.

[94] R. A. Wilson, 'The Local structure of the Lyons Group', preprint.

[95] J. R. Zemanek, 'Nilpotent Elements in Representation Rings', J. Algebra 19, 453-469, 1971.

[96] J. R. Zemanek, 'Nilpotent Elements in Representation Rings over Fields of Characteristic 2 ', J. Algebra 25, 534-553, 1973.

[97] V. M. Bondarenko and Yu.A. Drozd, 'The representation type of finite groups', Zap. Naučn. Sem. LOMI 57 (1977), 24-41.

[98] M. F. O'Reilly, 'On the modular representation algebra of a finite group', Ill. J. Math. 9 (1965), 261-276.

[99] H. R. Margolis, 'Spectra and the Steenrod Algebra', North-Holland 1983.

[100] M. Tezuka and N. Yagita , 'The varieties of the mod p cohomology rings of extra special p-groups for an odd prime p', Math. Proc. Camb. Phil. Soc. (1983), 94, 449-459.

Index

Lecture Notes in Mathematics

Edited by J.-M. Morel, F. Takens and B. Teissier

Editorial Policy
for the publication of monographs

1. Lecture Notes aim to report new developments in all areas of mathematics and their applications – quickly, informally and at a high level. Mathematical texts analysing new developments in modelling and numerical simulation are welcome.

   Monograph manuscripts should be reasonably self-contained and rounded off. Thus they may, and often will, present not only results of the author but also related work by other people. They may be based on specialised lecture courses. Furthermore, the manuscripts should provide sufficient motivation, examples and applications. This clearly distinguishes Lecture Notes from journal articles or technical reports which normally are very concise. Articles intended for a journal but too long to be accepted by most journals, usually do not have this „lecture notes" character. For similar reasons it is unusual for doctoral theses to be accepted for the Lecture Notes series, though habilitation theses may be appropriate.

2. Manuscripts should be submitted (preferably in duplicate) either to Springer's mathematics editorial in Heidelberg, or to one of the series editors (with a copy to Springer). In general, manuscripts will be sent out to 2 external referees for evaluation. If a decision cannot yet be reached on the basis of the first 2 reports, further referees may be contacted: The author will be informed of this. A final decision to publish can be made only on the basis of the complete manuscript, however a refereeing process leading to a preliminary decision can be based on a pre-final or incomplete manuscript. The strict minimum amount of material that will be considered should include a detailed outline describing the planned contents of each chapter, a bibliography and several sample chapters.

   Authors should be aware that incomplete or insufficiently close to final manuscripts almost always result in longer refereeing times and nevertheless unclear referees' recommendations, making further refereeing of a final draft necessary.

   Authors should also be aware that parallel submission of their manuscript to another publisher while under consideration for LNM will in general lead to immediate rejection.

3. Manuscripts should in general be submitted in English. Final manuscripts should contain at least 100 pages of mathematical text and should always include

   - a table of contents;
   - an informative introduction, with adequate motivation and perhaps some historical remarks: it should be accessible to a reader not intimately familiar with the topic treated;
   - a subject index: as a rule this is genuinely helpful for the reader.

   For evaluation purposes, manuscripts may be submitted in print or electronic form (print form is still preferred by most referees), in the latter case preferably as pdf- or zipped ps-files. Lecture Notes volumes are, as a rule, printed digitally from the authors' files. To ensure best results, authors are asked to use the LaTeX2e style files available from Springer's web-server at:

   ftp://ftp.springer.de/pub/tex/latex/mathegl/mono/ (for monographs) and

   ftp://ftp.springer.de/pub/tex/latex/mathegl/mult/ (for summer schools/tutorials).

   Additional technical instructions, if necessary, are available on request from lnm@springer.com.

4. Careful preparation of the manuscripts will help keep production time short besides ensuring satisfactory appearance of the finished book in print and online. After acceptance of the manuscript authors will be asked to prepare the final LaTeX source files (and also the corresponding dvi-, pdf- or zipped ps-file) together with the final printout made from these files. The LaTeX source files are essential for producing the full-text online version of the book (see http://www.springerlink.com/openurl.asp?genre=journal&issn=0075-8434 for the existing online volumes of LNM).

   The actual production of a Lecture Notes volume takes approximately 8 weeks.

5. Authors receive a total of 50 free copies of their volume, but no royalties. They are entitled to a discount of 33.3 % on the price of Springer books purchased for their personal use, if ordering directly from Springer.

6. Commitment to publish is made by letter of intent rather than by signing a formal contract. Springer-Verlag secures the copyright for each volume. Authors are free to reuse material contained in their LNM volumes in later publications: A brief written (or e-mail) request for formal permission is sufficient.

Addresses:

Professor J.-M. Morel, CMLA,
École Normale Supérieure de Cachan,
61 Avenue du Président Wilson, 94235 Cachan Cedex, France
E-mail: Jean-Michel.Morel@cmla.ens-cachan.fr

Professor F. Takens, Mathematisch Instituut,
Rijksuniversiteit Groningen, Postbus 800,
9700 AV Groningen, The Netherlands
E-mail: F.Takens@math.rug.nl

Professor B. Teissier, Institut Mathématique de Jussieu,
UMR 7586 du CNRS, Équipe "Géométrie et Dynamique",
175 rue du Chevaleret
75013 Paris, France
E-mail: teissier@math.jussieu.fr

For the "Mathematical Biosciences Subseries" of LNM:

Professor P. K. Maini, Center for Mathematical Biology,
Mathematical Institute, 24-29 St Giles,
Oxford OX1 3LP, UK
E-mail : maini@maths.ox.ac.uk

Springer, Mathematics Editorial, Tiergartenstr. 17,
69121 Heidelberg, Germany,
Tel.: +49 (6221) 487-8410
Fax: +49 (6221) 487-8355
E-mail: lnm@springer.com

9 783540 133896